T0298622

Advanced MOS Devices and their Circuit Applications

This text comprehensively discusses the advanced MOS devices and their circuit applications with reliability concerns. Further, an energy-efficient Tunnel FET-based circuit application will be investigated in terms of the output voltage, power efficiency, energy consumption, and performances using the device circuit co-design approach.

The book:

- Discusses advanced MOS devices and their circuit design for energy-efficient systems on chips (SoCs).
- Covers MOS devices, materials, and related semiconductor transistor technologies for the next-generation ultra-low-power applications.
- Examines the use of field-effect transistors for biosensing circuit applications and covers reliability design considerations and compact modeling of advanced low-power MOS transistors.
- Includes research problem statements with specifications and commercially available industry data in the appendix.
- Presents Verilog-A model-based simulations for circuit analysis.

The volume provides detailed discussions of DC and analog/RF characteristics, effects of trap-assisted tunneling (TAT) for reliability analysis, spacer-underlap engineering methodology, doping profile analysis, and work-function techniques. It further covers novel MOS devices including FinFET, Graphene field-effect transistor, Tunnel FETS, and Flash memory devices.

It will serve as an ideal design book for senior undergraduate students, graduate students, and academic researchers in the fields including electrical engineering, electronics and communication engineering, computer engineering, materials science, nanoscience, and nanotechnology.

Advanced MOS Devices and their Circuit Applications

Edited by
Ankur Beohar
Ribu Mathew
Abhishek Kumar Upadhyay
Santosh Kumar Vishvakarma

CRC Press
Taylor & Francis Group
Boca Raton London New York

CRC Press is an imprint of the
Taylor & Francis Group, an **informa** business

Cover image: Andrei Kuzmik/Shutterstock

First edition published 2024

by CRC Press
2385 NW Executive Center Dr Suite 320, Boca Raton, FL 33431

and by CRC Press
4 Park Square, Milton Park, Abingdon, Oxon, OX14 4RN

CRC Press is an imprint of Taylor & Francis Group, LLC

ISBN: 978-1-032-39285-1 (hbk)
ISBN: 978-1-032-67026-3 (pbk)
ISBN: 978-1-032-67027-0 (ebk)

DOI: 10.1201/9781032670270

Typeset in Sabon
by SPi Technologies India Pvt Ltd (Straive)

Contents

Editors

Dr. Ankur Beohar (Senior member IEEE) obtained a PhD degree in electrical engineering from the Indian Institute of Technology (IIT), Indore, MP, India, in 2018. After getting his PhD, he worked as a postdoctoral fellow in the Device Modeling Group, IISER, Bhopal, and then as a research scientist for one year under awarded Scientist Pool scheme of Council of Scientific and Industrial Research (CSIR), New Delhi. Currently, he is working as an assistant professor at Vellore Institute of Technology (VIT) Bhopal. He is an IEEE Senior Member and a Secretary of IEEE, Circuit and System Society, MP section, India. He completed his M.Tech degree in VLSI and Embedded System Design from MANIT Bhopal and B.Tech (Electronics) from RGPV University Bhopal in 2010 and 2005. He has a research and academic work experience of more than 13 years. He has a renowned research experience in the field of low-power device circuit design Memory Circuit Design and Reliability. His current research is related to new-generation innovative devices, such as optimization of gate all around (GAA)-Tunnel field effect transistor (TFET) with spacer engineering and its circuit applications. Currently, he is working in the research project sanctioned by the Science and Engineering Research Board (SERB) under the Teachers Associateship Research Excellence (TARE) scheme. Dr. Beohar has published more than 35 research publications in various peer-reviewed international conferences and SCI journals. Along with this, he has reviewed more than 100+ journal and conferences articles.

Dr. Ribu Mathew holds a doctorate degree in electronics engineering from Vellore Institute of Technology (VIT) University, Chennai Campus. A gold medallist in his post graduation, Dr. Mathew completed his MTech in VLSI design and BTech in electronics and communication engineering. In his doctoral research work, he has contributed in the field of design, modelling, and fabrication of NEMS technology piezoresistive readout-based nano cantilever sensors for chemical and biological sensing applications. In addition to the computational knowledge in industrial standard NEMS devices, he has gained experience in NEMS/IC layout tools and clean room fabrication technologies from CeNSE, IISc Bangalore. He has published several research papers in reputed international journals and conferences. His research areas include the design, modelling, and fabrication of MEMS/NEMS technology-based sensor and actuator systems, especially micro/nano cantilever and diaphragm-based devices, bio-MEMS, analog/RF IC design, SoC design, and device modeling. Currently he is working as an Associate Professor, MAHE, MANIPAL University, Karnataka.

Dr. Abhishek Kumar Upadhyay obtained a PhD in electrical engineering from the Indian Institute of Technology (IIT), Indore, MP, India, in 2019. After getting his PhD, he worked for one year as a postdoctoral fellow in the Model Group, Material to System Integration Laboratory, University of Bordeaux, France, and then as a staff scientist in the Chair of Electronics Devices and Integrated Circuits at Technische Universität Dresden, Germany, for two years. Currently he is working as an R&D rngineer in X-FAB GmbH, Dresden, Germany. He is the author of several research articles.

Professor Santosh Kumar Vishvakarma received the BSc in electronics from the University of Gorakhpur, Gorakhpur, in 1999, the MSc in electronics from Devi Ahilya Vishwavidyalaya, Indore, India, in 2001, the MTech in microelectronics from Punjab University, Chandigarh, India, in 2003, and the PhD in microelectronics and VLSI from the Department of Electronics and Communication Engineering, Indian Institute of

Technology Roorkee, India, in 2010. From 2009 to 2010, he was with University Graduate Center, Kjeller, Norway, as a postdoctoral fellow under European Union COMON project. Professor Vishvakarma is with the Department of Electrical Engineering, Indian Institute of Technology Indore, MP, India as a professor at IIT Indore. He is leading the Nanoscale Devices and VLSI Circuit and System Design (NSDCS) Laboratory since 2010. He is engaged with teaching and research in the areas of:

- Energy-efficient and reliable SRAM memory design
- Enhancing performance and configurable architecture for DNN accelerators
- SRAM based in-memory computing architecture for edge AI
- Reliable, secure design for IoT applications
- Design for reliability

He has supervised a total of seventeen PhD students, and currently six students are working with his group. He has authored or co-authored more than 175 research papers in peer-reviewed international journals and conferences. He was also granted 04 Indian Patent from IIT Indore and has thirteen sponsored research projects. He is a senior member of IEEE, professional member of VLSI Society of India, associate member of Institute of Nanotechnology, and life member of Indian Microelectronics Society (IMS), India.

Contributors

J. Ajayan
SR University
Hyderabad, Telangana, India

Sunanda Ambulkar
National Institute of Technology
Puducherry, India

A. V. Arun
Model Engineering College
Thrikakkara, Kochi

Ankur Beohar
VIT Bhopal
University Bhopal, India

Ajay Kumar Dadoria
Amity University
SIRT Bhopal
Bhopal, India

Narendra Kumar Garg
Amity University
SIRT Bhopal
Bhopal, India

Jobymol Jacob
College of Engineering Poonjar
Kottayam, India

Kavita Khare
Maulana Azad National Institute of
Technology (MANIT)
Bhopal, India

Vivek Singh Kushwah
Amity University
Gwalior, M.P, India

Ribu Mathew
Manipal Academy of Higher
Education (MAHE)
Manipal, Karnataka, India

Jitendra Kumar Mishra
NXPIndia Private limited
Pure, India

Uday Panwar
Sagar Institute of Research &
Technology (SIRT)
Bhopal, India

Neha Paras
National Institute of
Technology
Delhi, India

Siromani Balmukund Rahi
School of Information and
Communication Technology
Gautam Buddha University
Greater Noida, Uttar Pradesh,
India

Seema Rajput
VIT Bhopal University
Bhopal, India

Vancha Sharath Reddy
Department of Electronics and
    Communication Engineering
NIT Rourkela Odisha
Odisha, India

M. Sajeesh
Model Engineering College
Thrikakkara, Kochi, India

Soumya Sengupta
Department of Electronics and
    Communication Engineering
NIT Rourkela Odisha
Odisha, India

Trapti Sharma
School of Computing Science and
    Engineering
VIT Bhopal University
Bhopal, India

Billel Smaani
Centre Universitaire Abdelhafid
    Boussouf
Mila, Algeria

Shubham Tayal
Synopsys India Private Ltd.
Hyderabad, India

Laxman Raju Thoutam
Amrita Vishwa Vidyapeetham
Kochi, India

Ball Mukund Mani Tripathi
Velagapudi Ramakrishna
    Siddhartha Engineering
    College
Vijayawada, India

Abhishek Kumar Upadhyay
X-FAB Dresden Grenzstraße 28
Dresden, Germany

Sresta Valasa
National Institute of
    Technology
Warangal, India

Santosh Kumar Vishvakarma
Indian Institute of Technology
Indore, India

Arjun Singh Yadav
NIT Rourkela
Odisha, India

Preface

The field of metal oxide semiconductor (MOS) technology has experienced remarkable advancements in recent years, revolutionizing the electronics industry and enabling the development of smaller, more efficient devices. It is with great pleasure that we present this comprehensive book, *Advanced MOS Devices and Their Circuit Applications*, which delves into the intriguing realm of MOS device physics and explores its implications for circuit design.

This book serves as a valuable resource for semiconductor engineers, researchers, and students seeking a deeper understanding of MOS technology development. We have taken a unique approach, emphasizing the physical description, modeling, and technological implications of MOS devices, rather than focusing solely on formal aspects of device theory. By doing so, we aim to provide practical insights that can be readily applied in the real-world challenges faced by industry professional.

In this book, we dedicate significant attention to the critical issue of hot-carrier effects, investigating the engineering aspects of this problem and its impact on MOS devices. Additionally, we explore emerging low-temperature MOS technology, a rapidly evolving area that holds great promise for future advancements. Furthermore, we address the challenge of latch-up in scaled MOS circuits, shedding light on this complex phenomenon and its implications for circuit reliability.

To aid in comprehension and reinforce learning, each chapter is accompanied by a series of thought-provoking questions. In response to the feedback received since the publication of our previous book, *Fundamentals of Semiconductors: Physics and Materials Properties*, we have included solutions to selected problems in this new edition. While we recognize the value of problem-solving as an integral part of the learning process, we understand that some questions may be particularly challenging. Therefore, we have endeavored to provide additional study assistance by offering solutions to a subset of problems, leaving plenty of opportunities for instructors to assign problems and students to test their understanding of the material.

It is important to note that although this volume provides solutions to selected problems, there are often multiple approaches to solving a given problem. We encourage readers to explore alternative methods and think creatively, as this fosters a deeper understanding of the subject matter.

We extend our sincere gratitude to the researchers, engineers, and students who have contributed to the ever-evolving field of MOS technology. Their dedication and tireless efforts have paved the way for the remarkable progress we witness today. We hope that *Advanced MOS Devices and Their Circuit Applications* will serve as a valuable resource in your journey to unlock the potential of MOS technology.

Ankur Beohar

Acknowledgment

The authors would like to acknowledge the SERB TARE GRANT Project no. TAR/2022/000406, Govt. of India and VIT Bhopal University, Kothrikalan, Sehore-466114 for technical and financial support.

Chapter 1

An overview of DC/RF performance of nanosheet field effect transistor for future low-power applications

A. V. Arun and M. Sajeesh
Model Engineering College, Thrikakkara, India

Jobymol Jacob
College of Engineering Poonjar, Kottayam, India

J. Ajayan
SR University, Hyderabad, India

1.1 INTRODUCTION

The integration of billions of transistors in modern-day processors creates a different type of challenges for the semiconductor industry. The package density of chips should be improved to increase the speed of processing in modern-day gadgets. The solution to this problem is to scale down the transistor dimensions without degradation in its performance. The modern semiconductor industry has already developed 7-nm ICs, and efforts are ongoing to reduce the size further, to 5 nm. The existing 7-nm FinFET devices are already on the market and have succeeded in meeting the performance criteria [1–3]. At lower technology nodes, FinFET suffers from short channel effects and self-heating, and the patterning, compaction, and layout processes present certain challenges. The production cost also goes up as the transistors are scaled down. Another problem with FinFET devices' structure is that the bottom fingers are connected to the substrate, which leads to a high OFF current. This hurdle posed by FinFET can be rectified by implementing gate-all-around (GAA) FETs [4–6].

GAAFET exhibits better electrostatic control of the gate over the channel and is better immune to short-channel effects. This makes GAA nanowire (NW) FET a promising alternative to FinFET. The effective channel width to layout footprint ratio (J_F) is low for GAAFET. This leads to a lower switching speed in digital applications. Stacking of NW FETs in a vertical manner can increase the weff–LFP ratio. But the vertical stacking imposes high parasitic capacitance in the device, which yet again reduces the switching speed in digital circuits. When compared with gate-all-around NW FET and FinFET, nanosheet (NS) FET [7–9] gives high J_F ratio and lower parasitic capacitance.

DOI: 10.1201/9781032670270-1

1

This leads to high ON current in NS FET. The high-performance computing requirements of consumers are linked to the scaling-down of transistors for better characteristics. NSFETs are compatible with alternate materials, namely InGaAS, Ge, InSb, InAs, SiGe, and GeSn. This presents a wide variety of options to further improve the performance of the device. NS FETs are expected to emerge as the leading contender among the emerging semiconductor devices.

1.2 IMPACT OF WIDTH AND THICKNESS ON DEVICE PERFORMANCE

GAA NS-FET was first developed by IBM in 2015 [1]. The fabrication of horizontally stacked GAA NS-FET was done in 2017 [2]. This device has all the prospects to take the place of FinFET technology. The main advantage of this type of device is its high Weff/LFP ratio. The device fabricated in 2017 has a length of gate (L_g) equal to 12 nm. Contacted PolyPitch (CPP) is the minimum distance between two parallel Poly. The CPP of GAA NS-FET with horizontal stacking is 44/48 nm [2]. The 3D schematic of FinFET, stacked NW FET, and vertically stacked ND FET is shown in Figure 1.1. In FinFET, three fins cover the gate as shown in Figure 1.1(a).

This restricts the electrical control of the gate over the channel. The aggressive scaling-down of FinFET dimensions reduces the contact area and fin depopulation and increases the leakage current, which affects the DC/RF performance. In GAA NW-FET shown in Figure 1.1(b), horizontal and vertical stacking of nanowires is done to increase the ON current. The reduced effective channel width is the only limitation is NW-FET. Fin quantization and Fin pitch will not affect NS-FET, hence the width of the nanosheet (W_{NS}) can be optimized to achieve high drive current [3]. When the technology

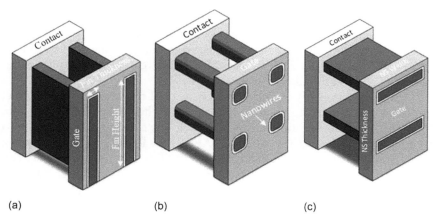

(a)   (b)   (c)

Figure 1.1 3-D schematic of (a) FinFET, (b) Stacked Nanowire-FET, (c) Vertically Stacked Nanosheet-FET.

node is scaled down beyond 7 nm, quantum mechanical effects should be also considered. The ballistic current is found to be decreasing with the channel width of all the devices, namely FinFET, GAA NW-FET, and GAA NS-FET.

The ballistic current [4] is given as

$$I_{ballistic} = qN_{inv}V_{inj} \tag{1.1}$$

where q is the electron charge, N_{inv} is the charge density of the inversion layer, and V_{inj} is the injected velocity.

The injected velocity in p-type SiGe NS-FET depends on WNS. As W_{NS} reduces, V_{inj} increases considerably. But in n-type Si NS-FET, V_{inj} is independent of W_{NS}. The low thickness of nanosheet reduces the subthreshold swing (SS) and offers better immunity to drain induced barrier lowering (DIBL). This is due to the improvement in gate control over the nanosheets. A P-channel and N-channel Ge-based GAA NSFET was first reported by Chun-Linchu et al. [5] in 2018. The Ge material has better charge mobility when compared with Si and SiGE materials. The fabrication of these devices can be done using low-pressure chemical vapor deposition (LP-CVD). A vertically stacked p-type GeSn-based GAA NS-FET was fabricated with a CVD process, as reported by Yu-Shiang Huang et al. [6]. Recently, GeSn emerged as a channel material due to its better hole mobility when compared with Ge. The effective mass of the holes gets reduced when Sn is incorporated into Ge. This causes an increase in the hole mobility. Compared with FinFET [7], NS-FET has higher ON current and lower RC delay at 7-nm technology mode. The RC delay, often called intrinsic delay [8], is given as

$$RC = C_{gg}V_{DD}/I_{ON} \tag{1.2}$$

where V_{DD} is the supply voltage and C_{gg} is the gate capacitance. Thus NS-FET shows better performance when compared with FinFET when dimensions are scaled down beyond 7 nm. This facilitates the application of the device in low-power analog or digital integrated circuits [9–14].

The fabrication procedure for defining the channel layer using selective sacrificial layer etching is unique to vertically stacked NS-FET when compared with FinFET. In vertically stacked NS-FET, the gap between nanosheets is filled by a sacrificial layer. SiGe is a popular sacrificial layer material. The etching techniques commonly used for removing the SiGe layer are dry etching [11] and chemical etching [12]. The problem with these techniques is that the selectivity is low, and hence there is a chance of removing a portion of the nanosheet. This leads to a reduction in nanosheet thickness, which leads to performance degradation. Different W_{NS} with vertically stacked nanosheets creates additional issues.

Aruna Kumari et al. [13] proposed the design of GAA NSFET with two vertical stacks, and its analog/RF performance was discussed. HfO_2 is used as the gate stack material to suppress the effect of gate leakage. Figure 1.2 shows the 2D and 3D cross-sections of GAA NSFET. A 16-nm gate is considered in the design. In this design, nanosheet width is denoted as N_W and nanosheet thickness is denoted as N_T. The transfer characteristics of GAA NSFET with a fixed gate length of 16 nm and a nanosheet 5 thickness of 5 nm are shown in Figure 1.3. The variation of ON current and OFF current with nanosheet width is shown in Figure 1.4.

For a fixed $V_{GS} = V_{DS} = 0.7$ V, the ON current of GAA NS FET is analyzed for various values of N_T and N_W as shown in Figure 1.5(a). As the device dimension increases, the ON current also increases proportionally [14]. The highest

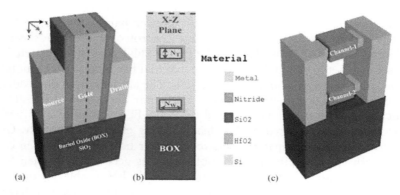

Figure 1.2 (a) 3D view of NS FET [13]; (b) 2D cross-section of NS FET with nanosheet thickness and width marked [13]; (c) 3D view of NS FET with two distinct channels [13].

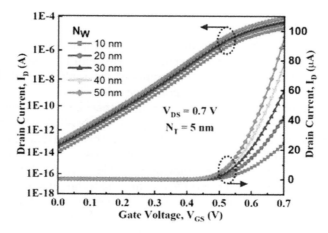

Figure 1.3 NS FET transfer characteristics with varying device dimensions [13].

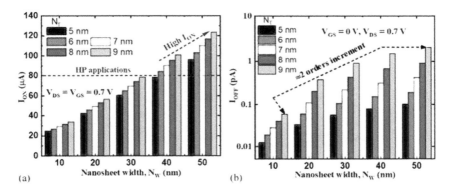

Figure 1.4 (a) ON current with varying nanosheet width and thickness; (b) OFF current with varying nanosheet width and thickness [13].

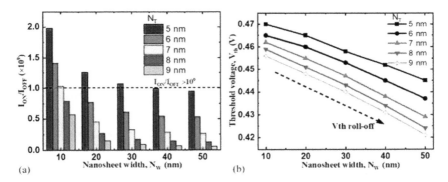

Figure 1.5 (a) ON/OFF current ratio with varying nanosheet width and thickness; (b) threshold voltage with varying nanosheet width and thickness [13].

ON current is obtained for N_T = 5 nm. Similarly, OFF current is analyzed in Figure 1.5(b) for various values of N_T and N_W. Like ON current, OFF currents also rise with the device dimensions due to the increase in leakage effect. As the nanosheet thickness reduces, ON current decreases slightly and the carrier mobility in the channel gets degraded. I_{ON}/I_{OFF} ratio is analyzed for various values of N_T and N_W as shown in Figure 1.6(a). As N_W increases, I_{ON}/I_{OFF} ratio decreases due to the increase in OFF current. As N_W increases, the leakage current gets incremented, and hence threshold voltage gets reduced as shown in Figure 1.6(b). The variation of SS and DIBL with N_T and N_W is shown in Figure 1.7(a) and (b) respectively. The minimum SS of 62.5 mV/decade is obtained with N_W = 10 nm and N_T = 5 nm. As the N_T and N_W is increased, DIBL also rises.

Hei Wong et al. [15] did a performance analysis on the impact of the channel width on three devices: FinFET, vertically stacked nanosheet FET, and vertically stacked nanowire FET. Figure 1.7 illustrates the device

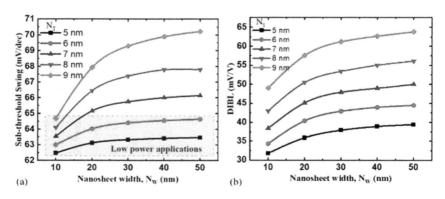

(a)

(b)

Figure 1.6 (a) SS with varying nanosheet width and thickness; (b) DIBL with varying nanosheet width and thickness [13].

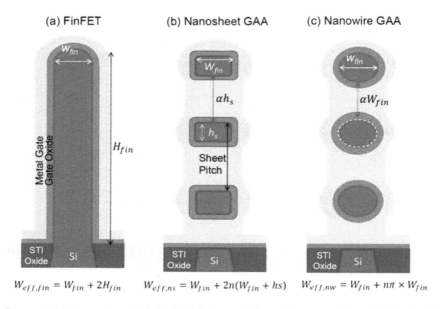

(a) FinFET   (b) Nanosheet GAA   (c) Nanowire GAA

$W_{eff,fin} = W_{fin} + 2H_{fin}$ $W_{eff,ns} = W_{fin} + 2n(W_{fin} + hs)$ $W_{eff,nw} = W_{fin} + n\pi \times W_{fin}$

Figure 1.7 Schematic of (a) FinFET, (b) vertical stacking nanosheet FET, and (c) vertical stacking nanowire FET [15].

cross-section of the above-mentioned three devices. The folding ratio with respect to channel width is defined as

$$\text{Folding Ratio} = \frac{W_{fin}}{W_{fin} + 2H_{fin}} \tag{1.3}$$

The folding ratio of the three devices are compared in Figure 1.8. The height of the fin is fixed at 80 nm. As the fin width increases from 5 to 15 nm,

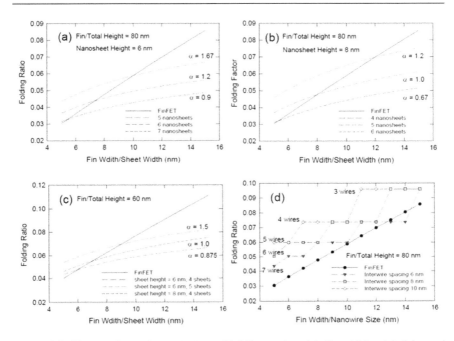

Figure 1.8 Plot to show the variation of folding ratio with fin width with (a) total height of 80 nm and sheet height of 6 nm for different α values, (b) total height of 80 nm and sheet height of 8 nm for different α values, (c) total height of 60 nm for different sheet height and α values, and (d) total height of 80 nm for different number of nanowires [15].

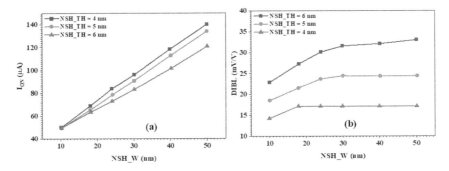

Figure 1.9 Plot showing the variation of NSH_W with (a) ON current and (b) change in DIBL effect of NS-FETs for different values of NSH_TH [16].

the folding ratio is found to be 0.03–0.087. If the sheet height is h_s, then the gap between the sheets is designed to be αh_s. From Figure 1.8(a), FinFET exhibits better folding ratio only when the fin width is greater than 10 nm. Nanosheet height of 8 nm is taken for the analysis in Figure 1.3(b) with the number of sheets varying from four to six. An α value of 0.67 is difficult to practically implement, as in this case the intersheet spacing becomes 5.3 nm. Vertically stacked nanosheet FET of 60-nm fin height is analyzed in Figure 1.8(c) for different α values. When fin width is above 7 or 9 nm, nanosheet FET shows better performance. A vertically stacked nanowire FET is analyzed in Figure 1.8(d). For a 14-nm wire, four wires can be stacked with a gap of 6 nm to get better performance than FinFET.

In vertically stacked NS-FET, the change in ON current with respect to thickness of nanosheet (NSH_TH) and width of the nanosheet (NSH_W) is shown in Figure 1.9(a).

As the nanosheet width increases, the ON current increases, and on the other hand a thinner nanosheet has high ON current. The OFF current also reduces if the thickness of the nanosheet is reduced. The increase in NSH_TH decreases the energy barrier in conduction band of the channel and OFF current is enhanced. DIBL effect also depends on the width and thickness of the nanosheet. As shown in Figure 1.9(b), an increase in NSH_TH and NSH_W causes a rise in DIBL. In Figure 1.10(a), the variation of I_{ON}/I_{OFF} ratio with nanosheet width and thickness is plotted.

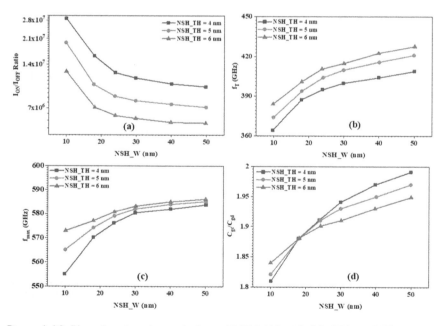

Figure 1.10   Plot showing the variation of NSH_W with (a) ON to OFF current ratio, (b) cutoff frequency (f_T), (c) maximum frequency of oscillation (f_{max}), and (d) gate to source capacitance to gate to drain capacitance ratio in vertically stacked GAA NS-FET for different values of NSH_TH [16].

It can be observed that thinner and wider channels are required for getting a high ON to OFF current ratio. The major AC parameters are cutoff frequency (f_T) and maximum frequency of operation (f_{max}). In vertically stacked GAA NS-FET, these parameters increase with the width of the nanosheet. This variation is shown in Figure 1.10(b) and (c). Therefore, the increase in nanosheet thickness improves f_{max} and f_T. This is due to an increase in gate-to-source capacitance (C_{gs}/C_{gd}) with the increase in nanosheet thickness. This observation can be seen in Figure 1.10(d). A thicker nanosheet leads to a high drain conductance and lowers the value of maximum intrinsic gain [16].

1.3 EFFECT OF SUBSTRATE MATERIALS

V Jegadheeeshan et al. [16] reported the impact of silicon-on-insulator (SOI), super-steep retrograde-silicon (SSR-Si), and punch-through stopper-silicon (PTs-Si) wafers on the overall performance of vertically stacked NS-FET. As the carrier mobility varies with different substrate materials, the electrical characteristics change with the type of substrate material used. Highest electron mobility is observed in SOI nanosheet, as the dopants cannot enter from wafer to the channel. The lowest electron mobility is observed in PTS-Si substrate as the dopants diffuse from wafer into channel during annealing. The electron mobility is degraded as Coulomb effects come into action because of the intrusion of dopants from substrate toward channel. As seen in Figure 1.11(a), SOI-based NS-FET provides higher transconductance when compared with PTS-Si and SSR-Si.

The drain current is also highest in SOI-based NS-FET, and because of the non-formation of parasitic channel, the OFF current will be minimal in this type of devices. As NSH_W increases, SOI-based NS-FETs exhibit higher

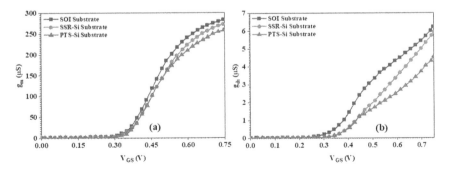

Figure 1.11 Plot showing the variation of (a) transconductance (g_m) and (b) drain conductance (g_{ds}) with input voltage on different substrates like SOI, SSR-Si, and PTS-Si in vertically stacked GAA NS-FET [16].

drain conductance (g_{ds}). This can be observed in Figure 1.11(b), as NSH_W increases the energy barrier in parasitic channel lowers.

1.4 EFFECT OF PARASITIC CHANNEL HEIGHT

The channel that exists below the vertically stacked nanosheet channels is known as parasitic channel. DC characteristics of GAA NS-FETs get affected by the height of parasitic channel (H_{ch}). Yuncho Choi et al. reported that DIBL and ON to OFF current ratio is better with an increase in H_{ch}. On the other hand, SS decreases with an increase in H_{ch}. The transfer characteristic is plotted in Figure 1.12(a) and drain current increases as H_{ch} increases. The variation of SS, DIBL and I_{ON}/I_{OFF} with H_{ch} is observed in Figure 1.12(b), 1.12(c), and 1.12(d), respectively.

The doping in the ground plane minimizes the impact of the presence of parasitic channel present in vertically stacked GAA NS-FETs. The punch through effect is minimized and threshold voltage can also be increased. The RC delay due to gate capacitance also increases with H_{ch} in GAA NS-FET. As Hch and ground plane doping increases, DIBL and SS of vertically stacked NS-FET can be decreased. A large value of H_{ch} also results in the increase of I_{ON}/I_{OFF} ratio.

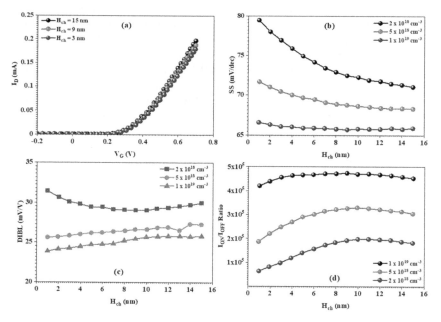

Figure 1.12 (a) Transfer characteristics for different values of H_{ch}, (b) subthreshold swing, (c) DIBL, and (d) ON current to OFF current ratio for different values of H_{ch} invertically stacked GAA NS-FET [17].

1.5 EFFECT OF SCALING DOWN THE GATE LENGTH

The reliability and performance of multi-gate transistors are seriously affected by self-heating effects. This effect occurs due to the scaling-down of gate lengths [18–21]. As L_G is scaled down in NS-FET, several short-channel effects (SCF) become significant. The change in electron velocity with transport direction for different devices like NS-FET, NW-FET, and FinFET is shown in Figure 1.13(a). Due to better sub-band occupation and gate control on NSchannel, GAA NS-FET exhibits highest electron velocity when compared with NW-FET and FinFET.

The characteristics of SS V_T and ON to OFF current ratio with L_G is shown in 1.13(b) to (d), respectively. SS improves in all devices with the scaling-down of L_G. The lowest value of SS can be observed in NW-FET. Furthermore, SS decreases with an increase in source/drain doping density ($N_{S/D}$). A higher value of source/drain doping leads to a low-threshold voltage and ON-OFF current ratio. GAA NS-FET shows better I_{ON}/I_{OFF} ratio when comparing with other devices' fixed device parameters. As mentioned before, V_T is reduced as L_G is scaled down, which results in increased OFF current. As W_{NS} is effectively larger and within low parasitic capacitance offered by

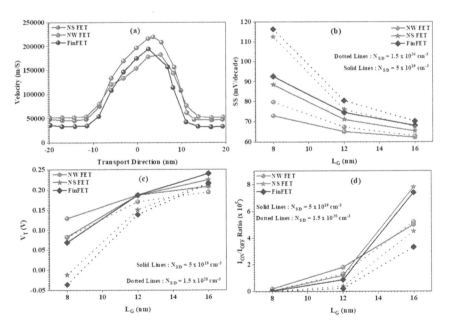

Figure 1.13 (a) The variation of electron velocity with respect to carrier transport direction for different devices. (b) The plot for showing the impact on SS when gate length is scaled down. (c) The plot for showing the impact on threshold voltage when gate length is scaled down. (d) The plot for showing the impact on ON to OFF current ratio when gate length is scaled down [22].

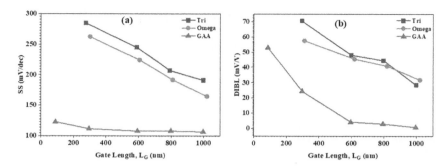

Figure 1.14 The variation of (a) SS and (b) DIBL with respect to gate length for different devices [23].

GAA NS-FET the drive current becomes larger. Due to this, volume of inversion improves under GAA NS-FET, which results in increased ON current. Meng Ju Tsai et al. [23] conducted experiments on trigate, omega gate, and gate-all-around structures to study the impact of scaling-down of gate lengths and observed that GAA based NS-FETs achieved reduced SCE. The GAA based structure provides low SS and DIBL when compared with other structures as shown in Figure 1.14(a) and (b), respectively.

These parameters justify that GAA NS-FETs offer better performance under down-scaling of gate length.

1.6 TEMPERATURE DEPENDANCE

Temperature is a significant parameter that affects DC/RF performance of NS-FET. A. S. Dahiya et al. [24] demonstrated the fabrication of ZnO nanosheet on gold-coated sapphire substrate. This is transformed into nanosheet-based source gated transistor (SGT). Figure 1.15 shows the cross-section of SGT with a platinum source and a titanium drain, which forms Schottky contact with the nanosheet.

The temperature-dependent transfer characteristics are shown in Figure 1.16(a) [25]. The ON current increases with an increase in temperature. The point to be noted is that there is no threshold shift as the temperature varies. The field mobility of the device is extracted as

$$\mu_{FE} = \frac{L}{W} \frac{g_m}{C_{ox}V_D} \tag{1.4}$$

where L and W are the channel length and width, respectively, g_m is the transconductance, and C_{ox} is the capacitance of the backgate. From Equation (1.4), peak value of mobility is obtained as 25 cm²/Vs at room temperature. The mobility obtained in the case of ohmic FET is 50 cm²/s. The mobility of SGT and of ohmic FET are compared in Figure 1.16(b) with respect to

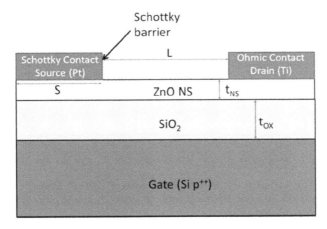

Figure 1.15 Cross-sectional view of nanosheet-based source gated transistor [24].

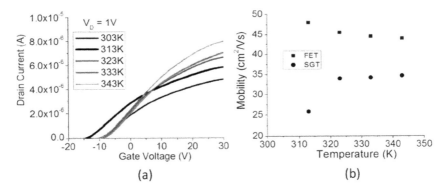

Figure 1.16 (a) Temperature-dependent transfer characteristics. (b) Comparison of mobility characteristics for SGT and ohmic FET [25].

temperature. The mobility is found to be increasing with the temperature in case of SGT, and it saturates for a particular temperature. In a FET device, the reverse is observed, as the temperature in mobility decreases. The conduction in SGT is controlled by the Schottky barrier at the source side, hence the transport is temperature dependent.

1.7 IMPACT OF GATE METAL WORK FUNCTION

The threshold voltage value can be modulated by the engineering of work function in gate metal and by changing the thickness of gate metal. JunSik Yoon et al. [26] conducted a study on n type and p type vertically stacked GAA-based NS-FET by varying the work function and thickness of gate metal. They conducted TCAD simulations by varying the gate metal work

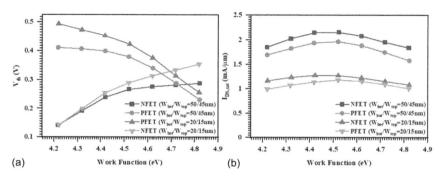

Figure 1.17 The variation of (a) threshold voltage and (b) drain to source saturation current with respect to work function for different device dimensions [26].

function within a range of 4.22–4.82 eV. The threshold voltage increases with an increase in gate metal work function in n-type NS-FET. The reverse characteristics are observed in p-type NS-FET. A decrease in threshold voltage leads to an increase in gate metal work function as observed in Figure 1.17a.

If the threshold voltage increases the OFF current and reduces the ON current, the ratio of the bottom-portion nanosheet width (W_{bot}) and top-portion nanosheet width (W_{top}) is a significant parameter that decides the value of threshold voltage and ON/OFF currents. A low threshold voltage is observed in NS-FETs having large W_{top} and W_{bot}. This impact is observed in Figure 1.12(a). The drive current is seen to be reduced when the work function is increased beyond 4.52 eV. This is because of the change in threshold voltage because of the change in the work function. The drive current of p-type NS-FET increases when W_{bot}/W_{top} ratio is small. This can be observed in Figure 1.17(b); when the W_{bot}/W_{top} ratio is 20/15, high drive is obtained. The nanosheet carrier density also varies when the work function of gate metal is changed. This variation leads to the shift of threshold voltage and ON/OFF current. A p type NS-FET having a gate metal work function of 4.22 eV will have a high value of hole density created at the center of the nanosheet. The top side and edges will have relatively low hole density. When the gate metal work function is increased beyond 4.22 eV, hole density near the edges increases, and only a small hole density is created at the bottom side. The trend in electron density in n-type NS-FET has the reverse characteristics when comparing with p type NS-FET [27–38].

1.8 RANDOM DOPANT FLUCTUATION

From Figure 1.18 it can be observed that the number of dopants in vertically stacked GAA NSFET [39] follows Poisson distribution and their position follows uniform distribution. The fluctuation in position and dopant count

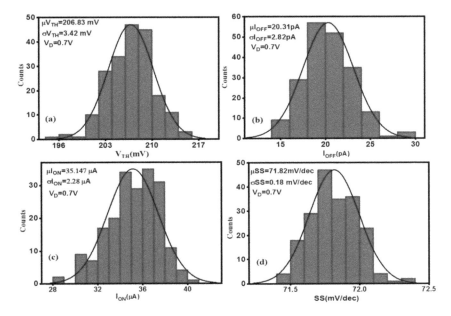

Figure 1.18 Distribution of dopant variation in source and drain regions with respect to (a) threshold voltage, (b) OFF current, (c) ON current, and (d) SS [39].

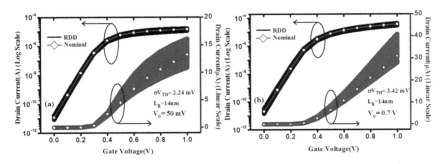

Figure 1.19 The plot showing the effect of random discrete dopants (RDD) on the transfer characteristics with (a) V_D = 50 mV (b) V_D = 0.7 V [39].

causes a change in transfer characteristics. Figure 1.19 shows the transfer characteristics of vertically stacked nanosheet GAA FET with a gate length of 14 nm. The random discrete dopants (RDD) cause a shift in threshold voltage and surface potential, hence the combined effect causes a change in transfer characteristics. This is due to the change in ON resistance in the source and drain regions because of RDD. As the device dimensions are reduced further, the RDD becomes more significant. The other DC/RF performance parameters such as threshold voltage, OFF current, ON current, and SS can be analyzed in Figure 1.18. The change in threshold voltage is given by

$$dV_{th}(y) = \frac{q}{\varepsilon_{Si}} \frac{dN_{sheet}(y)}{W_{NS}H_{NS}} \tag{1.5}$$

1.9 EMERGING RESEARCH IN THE FIELD OF NS-FET

As the technology advances toward a 3-nm node, GAA negative-capacitance NS-FET (GAA NC NS-FET) and TreeFET are found to be the most promising structures in their category. TreeFET has a fin-shaped inter-bridge channel that can increase the ON current of the device when compared with GAA NS-FET. By incorporating an inter-bridge channel with a fin, the effective channel area between source and drain is increased, which leads to an increase in carrier transport and, hence, improvement in ON current. The studies on the impact of inter-bridge height (H_{IB}) and inter-bridge width (W_{IB}) on the DC/RF performance of vertically stacked Ge-NS-FET was conducted by HungYu-Ye et al. [40]. The cross-sectional schematic of a TreeFET structure is shown in Figure 1.20.

The structure has the features of both FinFET and vertically stacked NS-FET. The device has an inter-bridge channel, oriented vertically, that connects the stacked nanosheets. The threshold voltage in nanosheet and inter-bridge channel should have an almost identical value to attain a high ON current in TreeFET. When the value of H_{IB} approaches 7.5 nm, the

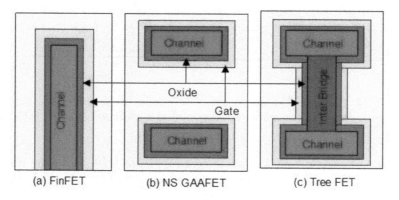

(a) FinFET   (b) NS GAAFET   (c) Tree FET

Figure 1.20 Two-dimensional schematic of (a) FinFET, (b) NS GAA FET, and (c) TreeFET [40].

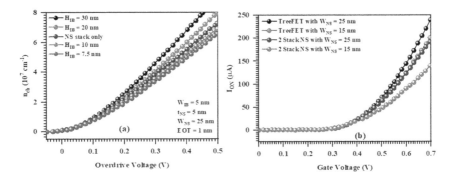

Figure 1.21 (a) The plot showing the variation of channel charge density with overdrive voltage for different inter-bridge height in TreeFET. (b) The plot showing the variation of ON current with gate voltage for different values of nanosheet width in TreeFET [40].

channel density (n_{ch}) is found to be very low when compared with the NS-FET structure without an inter-bridge. The inter bridge channel requires a minimum height of 10 nm to achieve sufficient n_{ch}, thereby increasing the ON current. The variation of n_{ch} with overdrive voltage for different values of HIB is shown in Figure 1.21(a).

As H_{IB} increases, n_{ch} increases; when H_{IB} approaches 10 nm, TreeFET offers better n_{ch} than stacked NS-FET, which thereby improves the ON current. The point to be noted is that H_{IB} does not have an impact on carrier density in a nanosheet channel, but it improves the charge density in an inter-bridge channel and thereby improving the ON current. The transfer characteristic of Tree FET is shown in Figure 1.21(b) for various values of W_{NS}. As seen in the previous sections, with the increase in W_{NS}, ON current improves.

Voltage scaling and sub-60 mV/decade SS can be achieved by incorporating negative capacitance (NC) effects in GAA NS-FET. Fahinul Sakch et al. [41] studied the DC capacitance of Si-based GAA NC NS-FET with 2D device simulations. A metal ferroelectric metal insulator (MFMIS) was stacked in the gate of GAA NS-FET to induce negative capacitance in the device. There was an improvement of 9% in SS when compared with NC NW FET and 38% in NC FinFET. The voltage across the ferroelectric layer (V_{FE}) changes with two parameters, namely ferroelectric layer thickness (T_{FE}) and charge on the gate (Q_G). The variation of I_{ON} and SS with respect to TFE is shown in Figure 1.22(a) and (b). I_{ON} increases with increase in T_{FE} and SS decreases with T_{FE}. The study was conducted with NW-FET, NS-FET, and FinFET. Hence a high value of T_{FE} is desirable to achieve a good DC/RF performance. NC NS-FET is found to be better performing when compared

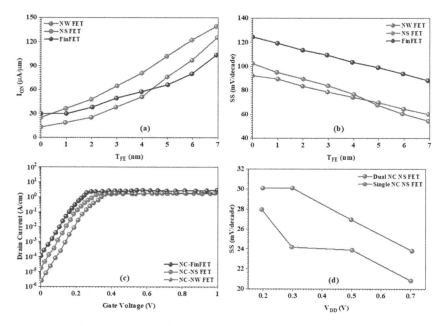

Figure 1.22 (a) The plot showing the variation of ON current with T_{FE} for different devices. (b) The plot showing the variation of SS with T_{FE} for different devices. (c) The transfer characteristics of different devices. (d) The variation of SS with supply voltage for different stackings in NC NS-FETs [41].

with other devices. The transfer characteristics of all the related devices are shown in Figure 1.22(c).

An analysis of the state-of-the-art NS-FET is illustrated in Table 1.1. NS-FETs, due to their scalability features without degrading the DC performance, make a viable alternative to FinFETs [42–63].

At the 3-nm technology node, NS-FETs are found to be providing the best performance. Various experiments demonstrate that at a 7-nm technology node, Si-based NS-FET exhibits high ON current and transconductance with negligible SCE when compared with FinFET.

Dynamic Random Access Memories (DRAM) and flash memories can be implemented onto a single NS-FET, as it has a stacked structure. This feature makes NS-FET extremely attractive for analog and digital circuits and for silicon-on-chip (SoC) applications. NS-FET having higher W_{NS} offers lower RC delayed high I_{ON}, g_m, and H. This makes the device suitable for low-power architectures. The conventional SI and Ge can be replaced with III-V elements such as InGaAS, InAs, InSb, and GAN in NS-FET to improve the ON current. High-gate dielectric materials can also be incorporated to achieve better performance when the device is scaled down to the 3-nm technology node.

Table 1.1 Comparison of DC performance of different NS FETs

Ref.	L_G (nm)	W_{NS} (nm)	Type of channel	Channel material	I_{ON} ($\mu A/\mu m$)	I_{ON}/I_{OFF} ratio	SS (mV/dec)
[5]	80	90	N	Ge	1510	10^5	140
[5]	80	90	P	Ge	1650	10^4	130
[2]	12	15	N	Si	–	10^6	75
[2]	12	15	P	Si	–	10^6	85
[63]	200	20–100	N	InGaAs	1350	–	–
[6]	60	18.7	P	GeSn	1975	7×10^4	108
[24]	75	60–100	N	SiGe	>500	5×10^8	150
[13]	14	25	N	Si	290.5	–	68.10
[13]	14	50	N	Si	283.7	–	70.01
[13]	14	75	N	Si	283.1	–	71.04
[16]	7	10	N	Si	68	2.6×10^7	71
[20]	2500	1500	N	ZnO	–	10^8	400
[23]	1000	1276	P	Si	–	7.5×10^6	187
[23]	1000	1300	P	Si	–	1.6×10^7	149
[23]	1000	4604	P	Si	–	2.03×10^8	100
[55]	60	135	P	SiGe	37.6	1.8×10^5	86
[22]	12	50	N	Si	699	1.2×10^5	71
[60]	12	40	P	Si	1130	1.13×10^5	79.8
[60]	12	40	N	Si	1410	1.41×10^5	73.9

1.10 CONCLUSION

This chapter analyzed the DC/RF performance of NS-FET under varying device parameters such as scaling of device dimensions, substrate materials, channel height, temperature, source/drain metal depth, and gate metal work function. The DC characteristics of NS-FET with different gate metals were analyzed. The crystal orientation of nanosheets was varied to study the change in DC/RF performance. NS-FET exhibits negligible SCE when moving towards 3-nm technology. The main problem with NS-FET structure is that quantum mechanical effects become significant at the 3-nm technology node. To rectify this issue, vertically structured GAA NS-FETs are being explored extensively by the research community. Furthermore, fin quantization is not a bottleneck in NS-FET technology. The width of the nanosheet does not depend on the fin quantization, and hence DC performance and power management can be optimized by tuning the nanosheet width. The memory devices can be implemented with a single nanosheet following stacked architecture. Thus it can be concluded that NS-FET offers a promising future for digital/analog circuit implementations.

REFERENCES

[1] S. Kim, M. Guillorn, I. Lauer, P. Oldiges, T. Hook and M. Na, Performance trade-offs in FinFET and gate-all-around device architectures for 7 nm-node and beyond, *2015 IEEE SOI-3D-Subthreshold Microelectronics Technology Unified Conference (S3S)*, Rohnert Park, CA, (2015), 1–3.

[2] N. Loubet, T. Hook, P. Montanini, C.-W. Yeung, S. Kanakasabapathy, M. Guillorn et al. Stacked nanosheet gate-all-around transistor to enable scaling beyond FinFET, *2017 Symposium on VLSI Technology*, Kyoto, (2017), T230–T231.

[3] D. Jang, D. Yakimets, G. Eneman, P. Schuddinck, M. G. Bardon, P. Raghavan, A. Spessot, D. Verkest, and A. Mocuta, Device Exploration of NanoSheet Transistors for Sub-7-nm Technology Node, *IEEE Transactions On Electron Devices*, 64 (2017), 2707–2713.

[4] M. S. Lundstrom and D. A. Antoniadis, Compact models and the physics of nanoscale FETs, *IEEE Transactions on Electron Devices*, 61 (2014), 225–233.

[5] C.-L. Chu, K. Wu, G.-L. Luo, B.-Y. Chen, S.-H. Chen, W.-F. Wu, and W.-K. Yeh, Stacked Ge-Nanosheet GAAFETs Fabricated by Ge/Si Multilayer Epitaxy, *IEEE Transactions on Electron Devices*, 39 (2018), 1133–1136.

[6] Y.-S. Huang, F.-L. Lu, Y.-J. Tsou, H.-Y. Ye, S.-Y. Lin, W.-H. Huang, and C. W. Liu, Vertically Stacked Strained 3-GeSn-Nanosheet pGAAFETs on Si Using GeSn/Ge CVD Epitaxial Growth and the Optimum Selective Channel Release Process, *IEEE Transactions on Electron Devices*, 39 (2018), 1274–1277.

[7] J. Yoon, J. Jeong, S. Lee and R. Baek, Systematic DC/AC Performance Benchmarking of Sub-7-nm Node FinFETs and Nanosheet FETs, *IEEE Journal of the Electron Devices Society*, 6 (2018), 942–947.

[8] E. J. Nowak, Ultimate CMOS ULSI performance, *Proceedings of IEEE International Electron Devices Meeting (IEDM)*, Washington, DC, (1993),115–118.

[9] J. Kim, J. Lee, J. Han and M. Meyyappan, Single-Event Transient in FinFETs and Nanosheet FETs," in *IEEE Electron Device Letters*, 39 (2018), 1840–1843.

[10] G. Hills, M. G. Bardon, G. Doornbos, D. Yakimets, P. Schuddinck, R. Baert, D. Jang, L. Mattii, S. M. Y. Sherazi, D. Rodopoulos, R. Ritzenthaler, C.-S. Lee, A. V-Y. Thean, I. Radu, A. Spessot, P. Debacker, F. Catthoor, P. Raghavan, M. M. Shulaker, H.-S. P. Wong, and S. Mitra, Understanding Energy Efficiency Benefits of Carbon Nanotube Field-Effect Transistors for Digital VLSI, *IEEE Transactions on Nanotechnology*, 17 (2018), 1259–1269.

[11] S. Borel, C. Arvet, J. Bilde, S. Harrison, and D. Louis, Isotropic etching of SiGe alloys with high selectivity to similar materials, *Microelectronic Engineering*, 73 (2004), 301–305.

[12] M. Orlowski, C. Ndoye, T. Liu, and M. Hudait, Si, SiGe, Ge, and III-V semiconductor nanomembranes and nanowires enabled by SiGe epitaxy, *ECS Transactions*, 33 (2010), 777–789.

[13] Kumari, N. Aruna, and P. Prithvi. "Device and circuit-level performance comparison of GAA nanosheet FET with varied geometrical parameters." *Microelectronics Journal* (2022): 105432.

[14] U. K. Das and T. K. Bhattacharyya, Opportunities in Device Scaling for 3-nm Node and Beyond: FinFET Versus GAA-FET Versus UFET, *IEEE Transactions on Electron Devices*, 67 (2020), 2633–2638.

[15] Wong, Hei, and Kuniyuki Kakushima. "On the Vertically Stacked Gate-All-Around Nanosheet and Nanowire Transistor Scaling beyond the 5 nm Technology Node." *Nanomaterials* 12.10 (2022): 1739.

[16] V. Jegadheesan, K. Sivasankaran, Aniruddha Konar, Impact of geometrical parameters and substrate on analog/RF performance of stacked nanosheet field effect transistor, *Materials Science in Semiconductor Processing*, 93 (2019), 188–195.

[17] Y. Choi, K. Lee, K. Y. Kim, S. Kim, J. Lee, R. Lee, H.-M. Kim, Y. S. Song, S. Kim, J.-H. Lee, B.-G. Park, Simulation of the effect of parasitic channel height on characteristics of stacked gate-all-around nanosheet FET, *Solid-State Electronics*, 164 (2020), 107686.

[18] W. Chen, L. Cai, K. Wang, X. Zhang, X. Liu, G. Du, Statistical simulation of selfheating induced variability and reliability with application to Nanosheet-FETs based SRAM, *Microelectronics Reliability*, 98 (2019), 63–68.

[19] G. Darbandy, S. Mothes, M. Schröter, A. Kloes, M. Claus, Performance analysis of parallel array of nanowires and a nanosheet in SG, DG and GAA FETs, *Solid-State Electronics*, 162 (2019), 107641.

[20] A. S. Dahiya, R.A. Sporea, G. P. Vittrant, D. Alquier, Stability evaluation of ZnO nanosheet based source-gated transistors, *Scientific Reports*, 9 (2019), 2979.

[21] J. Yoon, J. Jeong, S. Lee and R. Baek, Punch-Through-Stopper Free Nanosheet FETs With Crescent Inner-Spacer and Isolated Source/Drain, *IEEE Access*, 7 (2019), 3859338596.

[22] D. Nagy, G. Espiñeira, G. Indalecio, A. J. García-Loureiro, K. Kalna and N. Seoane, Benchmarking of FinFET, Nanosheet, and Nanowire FET Architectures for Future Technology Nodes, *IEEE Access*, 8 (2020), 53196–53202.

[23] M.-J. Tsai, K.-H. Peng, C.-J. Sun, S.-C. Yan, C.-C. Hsu, Y.-R. Lin, Y.-H. Lin, Y.-C. Wu, Fabrication and Characterization of Stacked Poly-Si Nanosheet With Gate-All-Around and Multi-Gate Junctionless Field Effect Transistors, *IEEE Journal of the Electron Devices Society*, 7 (2019), 1133–1139.

[24] Dahiya, Abhishek Singh, et al. "Temperature dependence of charge transport in zinc oxide nanosheet source-gated transistors." *Thin Solid Films* 617 (2016): 114–119.

[25] S. Venkateswarlu and K. Nayak, Hetero-Interfacial Thermal Resistance Effects on Device Performance of Stacked Gate-All-Around Nanosheet FET, *IEEE Transactions on Electron Devices*, 67 (2020), 4493–4499.

[26] J. Yoon, J. Jeong, S. Lee and R. Baek, Multi-V_{th} Strategies of 7-nm node Nanosheet FETs With Limited Nanosheet Spacing, *IEEE Journal of the Electron Devices Society*, 6 (2018), 861–865.

[27] J. Zhang, T. Ando, C. W. Yeung, M. Wang, O. Kwon, R. Galatage, R Chao et al., High-k metal gate fundamental learning and multi-Vt options for stacked nanosheet gate-all-around transistor, *2017 IEEE International Electron Devices Meeting (IEDM)*, San Francisco, CA, 2017, pp. 22.1.1–22.1.4.

[28] S. Barraud, V. Lapras, B. Previtali, M. P. Samson, J. Lacord, S. Martinie, M.-A. Jaud, S. Athanasiou, F. Triozon et al., "Performance and design considerations for gate-allaround stacked-Nano Wires FETs," *2017 IEEE International Electron Devices Meeting (IEDM)*, San Francisco, CA, 2017, pp. 29.2.1–29.2.4.

[29] R. Huang, X. B. Jiang, S. F. Guo, P. P. Ren, P. Hao, Z. Q. Yu, Z. Zhang, Y. Y. Wang, R. S. Wang et al., Variability-and reliability-aware design for 16/14nm and beyond technology, *2017 IEEE International Electron Devices Meeting (IEDM)*, San Francisco, CA, 2017, pp. 12.4.1–12.4.4.

[30] S. Ramesh, T. Ivanov, V. Putcha, A. Alian, A. S. Hernandez, R. Rooyackers, E. Camerotto, A. Milenin, N. Pinna et al., Record performance Top-down In0.53Ga0.47As vertical nanowire FETs and vertical nanosheets, *2017 IEEE International Electron Devices Meeting (IEDM)*, San Francisco, CA, 2017, pp. 17.1.1–17.1.4.

[31] N. Yoshida, S. Hassan, W. Tang, Y. Yang, W. Zhang, S. C. Chen, L. Dong, H. Zhou, M. Jin, M. Okazaki, J. Park, et al., Highly conductive metal gate fill integration solution for extremely scaled RMG stack for 5 nm & beyond, *2017 IEEE International Electron Devices Meeting (IEDM)*, San Francisco, CA, 2017, pp. 22.2.1–22.2.4.

[32] X. He, J. Fronheiser, P. Zhao, Z. Hu, S. Uppal, X. Wu, Y. Hu, R. Sporer, L. Qin, R. Krishnan, E. M. Bazizi, R. Carter et al., Impact of aggressive fin width scaling on FinFET device characteristics, *2017 IEEE International Electron Devices Meeting (IEDM)*, San Francisco, CA, 2017, pp. 20.2.1–20.2.4.

[33] Z. Krivokapic, U. Rana, R. Galatage, A. Razavieh, A. Aziz, J. Liu, J. Shi, H. J. Kim, R. Sporer, C. Serrao, A. Busquet et al., 14nm Ferroelectric FinFET technology with steep subthreshold slope for ultra low power applications, *2017 IEEE International Electron Devices Meeting (IEDM)*, San Francisco, CA, 2017, pp. 15.1.1–15.1.4.

[34] A. Vardi, L. Kong, W. Lu, X. Cai, X. Zhao, J. Grajal, Jesús A. del Alamo et al., Selfaligned InGaAs FinFETs with 5-nm fin-width and 5-nm gate-contact separation, *2017 IEEE International Electron Devices Meeting (IEDM)*, San Francisco, CA, 2017, pp. 17.6.1–17.6.4.

[35] X. Zhao, C. Heidelberger, E. A. Fitzgerald, W. Lu, A. Vardi and J. A. del Alamo, Sub10 nm diameter InGaAs vertical nanowire MOSFETs, *2017 IEEE International Electron Devices Meeting (IEDM)*, San Francisco, CA, 2017, pp. 17.2.1–17.2.4.

[36] O. Badami, F. Driussi, P. Palestri, L. Selmi and D. Esseni, Performance comparison for FinFETs, nanowire and stacked nanowires FETs: Focus on the influence of surface roughness and thermal effects, *2017 IEEE International Electron Devices Meeting (IEDM)*, San Francisco, CA, 2017, pp. 13.2.1–13.2.4.

[37] V. P. Hu, P. Chiu, A. B. Sachid and C. Hu, Negative capacitance enables FinFET and FDSOI scaling to 2 nm node, *2017 IEEE International Electron Devices Meeting (IEDM)*, San Francisco, CA, 2017, pp. 23.1.1–23.1.4.

[38] D. Yakimets, M. G. Bardon, D. Jang, P. Schuddinck, Y. Sherazi, P. Weckx, K. Miyaguchi, B. Parvais, P. Raghavan et al., Power aware FinFET and lateral nanosheet FET targeting for 3nm CMOS technology, *2017 IEEE International Electron Devices Meeting (IEDM)*, San Francisco, CA, 2017, pp. 20.4.1–20.4.4.

[39] Mohapatra, E., et al. "Design study of gate-all-around vertically stacked nanosheet FETs for sub-7nm nodes." *SN Applied Sciences* 3.5 (2021): 1–13.

[40] H.-Y. Ye and C. W. Liu, On-Current Enhancement in TreeFET by Combining Vertically Stacked Nanosheets and Interbridges, *IEEE Electron Device Letters*, 41 (2020), 1292–1295.

[41] F. I. Sakib, M. A. Hasan and M. Hossain, Exploration of Negative Capacitance in Gate All-Around Si Nanosheet Transistors, *IEEE Transactions on Electron Devices*, 67 (2020), 5236–5242.

[42] A. V. de Oliveira, A. Veloso, C. Claeys, N. Horiguchi and E. Simoen, Low–Frequency Noise in Vertically Stacked Si n–Channel Nanosheet FETs, *IEEE Electron Device Letters*, 41 (2020), 317–320.

[43] A. Dasgupta, S. S. Parihar, P. Kushwaha, H. Agarwal, M.-Y. Kao, S. Salahuddin, Y. S. Chauhan, and C. Hu, BSIM Compact Model of Quantum Confinement in Advanced Nanosheet FETs, *IEEE Transactions on Electron Devices*, 67 (2020), 730737.

[44] J. Yoon, J. Jeong, S. Lee and R. Baek, Sensitivity of Source/Drain Critical Dimension Variations for Sub-5-nm Node Fin and Nanosheet FETs, *IEEE Transactions on Electron Devices*, 67 (2020), 258–262.

[45] A. Oliveira, A. Veloso, C. Claeys, N. Horiguchi and E. Simoen, Low-Frequency Noise Assessment of Vertically Stacked Si n-Channel Nanosheet FETs With Different Metal Gates, *IEEE Transactions on Electron Devices*, 67 (2020), 4802–4807.

[46] C. K. Jha, P. Yogi, C. Gupta, A. Gupta, R. Vega and A. Dixit, Comparison of LER Induced Mismatch in NWFET and NSFET for 5-nm CMOS, *IEEE Journal of the Electron Devices Society*, 8 (2020), 1184–1192.

[47] C. K. Jha, C. Gupta, A. Gupta, R. A. Vega and A. Dixit, Impact of LER on Mismatch in Nanosheet Transistors for 5nm-CMOS, *4th IEEE Electron Devices Technology & Manufacturing Conference (EDTM)*, Penang, Malaysia, (2020), 1–4.

[48] V. Jegadheesan, K. Sivasankaran and A. Konar, Optimized Substrate for Improved Performance of Stacked Nanosheet Field-Effect Transistor, *IEEE Transactions on Electron Devices*, 67 (2020), 4079–4084.

[49] A. D. Gaidhane, G. Pahwa, A. Dasgupta, A. Verma and Y. S. Chauhan, Compact Modeling of Surface Potential, Drain Current and Terminal Charges in Negative Capacitance Nanosheet FET including Quasi-Ballistic Transport, *IEEE Journal of the Electron Devices Society*, 8 (2020), 1168–1176.

[50] P.-J. Sung, S.-W. Chang, K.-H. Kao, C.-T. Wu, C.-J. Su, T.-C. Cho, F.-K. Hsueh, W-H Lee, Y-J. Lee, and T-S. Chao, Fabrication of Vertically Stacked Nanosheet Junctionless Field-Effect Transistors and Applications for the CMOS and CFET Inverters, *IEEE Transactions on Electron Devices*, 67 (2020), 3504–3509.

[51] A. Dasgupta, S. S. Parihar, H. Agarwal, P. Kushwaha, Y. S. Chauhan and C. Hu, Compact Model for Geometry Dependent Mobility in Nanosheet FETs, *IEEE Electron Device Letters*, 41 (2020), 313–316.

[52] N. Gyanchandani, S. Pawar, P. Maheshwary, K. Nemade, Preparation of spin-tronically active ferromagnetic contacts based on Fe, Co and Ni Graphene nanosheets for SpinField Effect Transistor, *Materials Science and Engineering: B*, 261 (2020), 114772.

[53] Te-K. Chiang, Nanosheet FET: A new subthreshold current model caused by interface-trapped-charge and its application for evaluation of subthreshold logic gate, *Microelectronics Journal*, 104 (2020), 104893.

[54] A. Veloso, T. Huynh-Bao, P. Matagne, D. Jang, G. Eneman, N. Horiguchi, J. Ryckaert, Nanowire & nanosheet FETs for ultra-scaled, high-density logic and memory applications, *Solid-State Electronics*, 168 (2020), 107736.

[55] X. Yin, Y. Zhang, H. Zhu, G. L. Wang et al., Vertical Sandwich Gate-All-Around Field-Effect Transistors With Self-Aligned High-k Metal Gates and Small Effective Gate-Length Variation, *IEEE Electron Device Letters*, 41 (2020), 8–11.

[56] U. K. Das and T. K. Bhattacharyya, Opportunities in Device Scaling for 3-nm Node and Beyond: FinFET Versus GAA-FET Versus UFET, *IEEE Transactions on Electron Devices*, 67 (2020), 2633–2638.

[57] D.-I. Bae, B.-D. Choi, Short channels and mobility control of GAA multi stacked nanosheets through the perfect removal of SiGe and post treatment, *Electronics Letters*, 56 (2020), 400–402.

[58] Q. Huo, Z. Wu, W. Huang, X. Wang, G. Tang, J. Yao, Y. Liu et al., A Novel General Compact Model Approach for 7-nm Technology Node Circuit Optimization From Device Perspective and Beyond, *IEEE Journal of the Electron Devices Society*, 8 (2020), 295–301.

[59] J. Yoon, S. Lee, J. Lee, J. Jeong, H. Yun and R. Baek, Reduction of Process Variations for Sub-5-nm Node Fin and Nanosheet FETs Using Novel Process Scheme, *IEEE Transactions on Electron Devices*, 67 (2020), 2732–2737.

[60] J. Jeong, J. Yoon, S. Lee and R. Baek, Comprehensive Analysis of Source and Drain Recess Depth Variations on Silicon Nanosheet FETs for Sub 5-nm Node SoC Application, *IEEE Access*, 8 (2020), 35873–35881.

[61] V. Jegadheesan, K. Sivasankaran, A source/drain-on-insulator structure to improve the performance of stacked nanosheet field-effect transistors. *Journal of Computational Electronics*, 19 (2020), 1136–1143.

[62] D. Ryu, M. Kim, S. Kim, Y. Choi, J. Yu, J.-H. Lee, B.-G. Park et al., Design and Optimization of Triple-k Spacer Structure in Two-Stack Nanosheet FET From OFFState Leakage Perspective, *IEEE Transactions on Electron Devices*, 67 (2020), 13171322.

[63] J. J. Gu, X. W. Wang, J. Shao, A. T. Neal, M. J. Manfra, R. G. Gordon, and P. D. Ye, III-V gate-all-around nanowire MOSFET process technology: From 3D to 4D, *IEDM Technical Digest*, December (2012), 23.7.1–23.7.4.

Chapter 2

Device design and analysis of 3D SCwRD cylindrical (Cyl) gate-all-around (GAA) tunnel FET using split-channel and spacer engineering

Ankur Beohar and Seema Rajput
VIT Bhopal University, Bhopal, India

Kavita Khare
Maulana Azad National Institute of Technology (MANIT), Bhopal, India

Santosh Kumar Vishvakarma
Indian Institute of Technology, Indore, India

2.1 INTRODUCTION

As discussed in the previous chapter, the ongoing downsizing of the MOSFET in the nanometer regime resulted in severe short channel effects (SCEs), which encourage high drip and knowingly reduce the device performance in terms of DC characteristics. In this regard, Tunnel FET appears as a prominent novel device and an alternative to MOSFET for low-power applications.

Asymmetric source/drain doping engineering in Tunnel FET suppresses ambipolar conduction. However, Tunnel Field Effect Transistor (TFET) suffers from low ON current [1] due to a different conduction concept as compared to conventional MOSFET. This causes degradation in analog/RF performances and consequently limited utilization of Tunnel FET for low-power applications. It was also reported that leakage current is a dominating factor in the nanometer dimension [2]. Thus drain design optimization is an important concern for improving the device performance and its RF figures of merit without an increase in OFF current. In addition, insertion engineering significantly affects the TFET in terms of improved DC characteristics. Please note that insertions are basically insulator vigorous for isolation to prevent the carrier leakage over the gate edge. Based on the literature of the fabricated device [2], insertions prevent source/drain dopants from being implanted through any flush faceted regions.

Therefore, in this chapter, impacts of split channel with retrograde doping (SCwRD) gate-all-around (GAA)-TFET are analyzed for the improvement of the device performance. We investigate the examined devices in terms of

DC characteristics such as I_{ON}, I_{OFF}, SS, and I_{ON}/I_{OFF}. Additionally, we evaluate the influence of low insertion width on 3D Cyl GAA-n-channel TFET with the equivalent oxide thickness and compare the performance to that achieved with symmetrical insertion width. It chances out that the fringing field effect with low insertion width produces a high I_{ON} when compared with the device based on high insertion width [3–6].

2.2 DEVICE STRUCTURE AND ANALYSIS

3D Cyl GAA-Tunnel FET is purely a GAA p-i-n diode operating under a reverse bias as shown in Figures 2.1 and 2.2.

The examined device has following parameters; asymmetrical doping profile is used for source, channel, and drain region to make abrupt junction. P-type source ($1 \times 10^{19}/cm^3$), n-type drain ($5 \times 10^{18}/cm^3$), and p-type

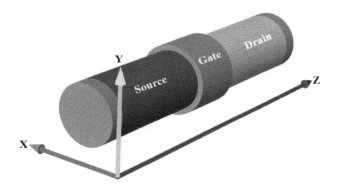

Figure 2.1 Three-dimensional view of Cyl SCwRD GAA Tunnel FET with low insertion width (LSW) and split channel. S, Source; D, drain; C, channel.

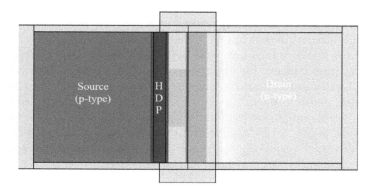

Figure 2.2 Cross-sectional views of SCwRD GAA-TFET with low insertion width (LSW) along the channel length direction.

channel region (10^{17}/cm^3) are shown in Figures 2.1 and 2.2 with scale of doping concentrations. Because of augmented oxide capacitance, HfO_2 is used as a gate dielectric as well as insertion dielectric (k = 21) with gate work function = 4.21 eV.

2.3 SIMULATION MODEL AND PARAMETERS

The pinpointing keyword is used in the MODELS proclamation to activate the Low Field Mobility Model method. This model aids in the specification of low field mobilizes that are temperature and doping related. This model is also defined by default for silicon at 300 K; to activate this model, use Lombardi CVT (Continuously Variable Transmission) Model. This paradigm prioritizes mobility in a way that no other mobility model does. Mathematician's rule is utilized in this model to combine the components involved with mobility that is affected by transverse field, temperature, and doping. Parallel Electric Field Mobility Model will be active by default when this model is activated. Model of Shockley-Read-Hall Recombination: This model is activated using the SRH parameter in the MODELS statement [7].

There's a few definable variables in the material statement, such as the electron and hole lifetime parameters TAUN0 and TAUP0. This model depicts electron and hole interaction within the device using the Shockley-Read-Hall recombine mechanism. Model of Auger Recombination: The MODELS statement may be used to activate this model by providing AUGER. The MATERIAL statement can use the coefficients augn and augp; Boltzmann Model: This is the default carrier statistics model. BOLTZMANN is specified in the MODELS statement to turn it on. This model uses Boltzmann statistics, as the name implies; Fermi-Dirac Model: This model uses Fermi-Dirac statistics. It frequently occurs in areas that are strongly doped yet have low carrier concentrations [8–10].

The device modeling specifications used in TCAD Synopsys simulation are as follows: Device parameters for SCwRD GAA-TFET: Oxide thickness (t_{ox}) = 2 nm, channel length (Lch) = 20 nm, p-type channel doping concentration of core region = 10^{15} cm^{-3} and outer region = 10^{14} cm^{-3}, source (p-type) = 10^{20} cm^{-3}, drain (n-type) = 5×10^{18} cm^{-3}, radi (r) = 7 nm, gate work function = 4.53 eV, extension length of source (Lextss) & drain (Lextdd) = 40 nm, spacer length (S_L) = 10 nm, length of suppressed drain (LSD) = 2 nm. The material used for drain area is GaAs. For the source, a low-k oxide (Silicon dioxide (SiO_2), k = 3.9) material is employed, while a high-k oxide (Hafnium Oxide (HfO_2), k = 22) substance spacer is utilized across the channel and drain areas. TCAD Synopsys was used to model and simulate the device [11–14]. The following are some of the usual models that have been used to imitate device efficiency when modeling a device that is comparable to the actual thing: Low-Field Mobility Model with Intensity Dependence: CONMOB is used in the MODELS statement to activate this model. For silicon and gallium arsenide alone, this model offers data for low field mobilities of electrons and holes at 300 K.

2.4 RESULTS AND DISCUSSION

In this work, the proposed device is split-channel with retrograde doping GAA-TFET. The devices include the HeO dielectric and SCwRD (split channel with retrograde doping). However, HeO shows the existence of two distinct materials oxide, at the source as well as a source-side channel with low-k dielectric (SiO_2, k = 3.9), which lowers the fringing fields for the specific area, resulting in a smaller tunnel barrier width and improved ON-current (I_{ON}), although the presence of a high-k dielectric substance (HfO_2, k = 22) at the drain and drain-side channel strengthens the tunnel cordon at the drain-to-channel junction, leading to a high OFF-current (I_{OFF}). Furthermore, for HeO topology, dual-substrate material source and gate area with Silicon-Germanium (low band-gap material, Eg = 0.78 eV) and drain region with silicon (high band gap material, Eg = 1.12 eV) has been integrated. Figure 2.1 represents a 3D view for SCwRD GAA-TFET device whereas Figure 2.2 depicts a cross-sectional view for the topologies of SCwRD GAA-TFET. SCRD was also employed in the channel region, with a smaller band gap substrate (SiGe) at the source-to-channel edge and a greater band gap substrate (Gallium Arsenide (GaAs)) at the drain-to-channel edge. Because the drain-side doping is more than the source, the I_{OFF} depreciates more, resulting in a modest improvement in the I_{ON} and I_{OFF}. These electrons begin to wander from the drain to the source as a result of the supply voltage, resulting in an increase in current density around the channel-drain area. As a result, the SCwRD GAA-TFET has an improved penetration performance. The RD was used to define a high doping zone (inner region) that contains 30% of the channel region, with the remaining portion being less doped than the inner region, as illustrated in Figure 2.2. SCE is reduced while I_{OFF} and SS are improved at the same time. By reducing the electric field at the drain side and allowing for a lower gate-drain capacitance, the HDP enhances ambipolar properties (Cgd).

With a voltage of Vgs = 1 V, the maximum ION = 1.78×10^{-6} A/m and the lowest IOFF = 4.12×10^{-18} A/m are achieved, as shown in Figure 2.4. Due to the existence of high-k, which obscures the flow of charges toward the drain side, there is a stoppage in the oxide level at the drain side. RD is used in conjunction with HDP in this study. The core section of the channel has a larger carrier concentration than the channel's perimeter due to RD, leading to a higher BTBT at the source–channel junction.

Figure 2.4 depicts the effect of channel splitting with retrograde doping concentration on band-gap energies. In SCwRD GAA-TFET, the energy band gap between the valance and conduction bands is found to be small. The energy of conduction band is lowered in the ON state when drain-side

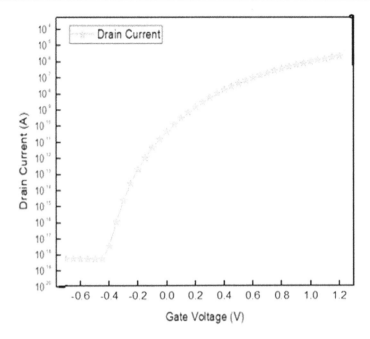

Figure 2.3 Transfer characteristics of Id vs Vgs for SCwRD GAA-TFET.

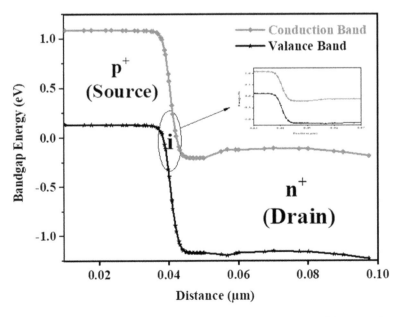

Figure 2.4 Energy band characteristics for SCwRD GAA-TFET (ON state).

channel doping increases, decreasing the band gap between conduction and valance band energy levels (Ec) (Ev). Electrons go from the valance band to the conduction band as a result of this. In the case of a SCwRD GAA-TFET, the electron conduction is larger, since the energy gap (Eg) is smaller. As a consequence, the quantum of electrons flowing from the valance band to the conduction band increases significantly. As a result, increased tunneling ensures that the device leaks less. The device's leakage condition improves as a result of this.

Figure 2.5 depicts the Cgs for SCwRD GAA-TFET. Cgs = 0.25 fF is given as the gate-source capacitance for SCwRD GAA-TFET. Only source-side intrinsic fringing field capacitance (C_{SIF}) and exterior fringing-field capacitance (C_{SOF}) may be utilized to estimate the value of Cgs in the case of gate-source capacitance.

Cgd for SCwRD GAA-TFET is depicted in Figure 2.6. Cgd is the device's miller capacitance, and in order to improve switching speed and analogue performance, the miller capacitance should be reduced. The miller capacitance for SCwRD GAA-TFET is found to be lower Cgd = 80 × 10⁻¹⁸ F. As indicated parasitic capacitances such as gate-drain overlap capacitance (C_{GDO}), inner drain-side fringing field capacitance (C_{DIF}), and outer drain-side fringing field capacitance (C_{DOF}) are effective to influence the analog/RF performance of the device

Figure 2.7 depicts the transconductance for SCwRD GAA-TFET. The value of gm = 58 μA/V is for SCwRD GAA-TFET. The variable transconductance (gm) is essential in determining a device's current driving potential.

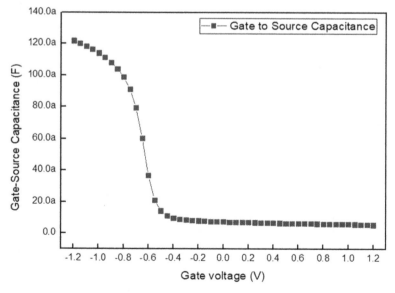

Figure 2.5 Gate source capacitance (Cgs) vs gate voltage (Vgs) for SCwRD GAATFET.

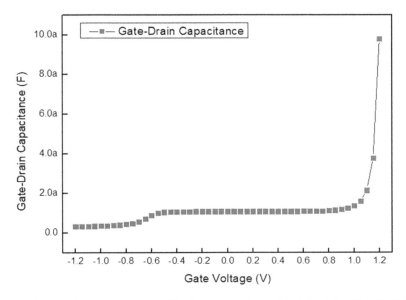

Figure 2.6 Gate drain capacitance (Cgd) vs gate voltage (Vgs) for SCwRD GAATFET.

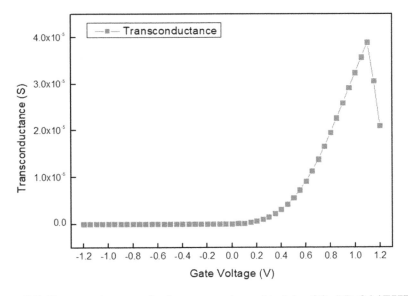

Figure 2.7 Transconductance (gm) vs gate voltage (Vgs) for SCwRD GAATFET.

The gm specifies how successfully a low Vgs at the transistor gate is converted into a drain current signal. In respect to output resistance, the intrinsic transistor gain reflects how often a transistor acts as a controlled current source.

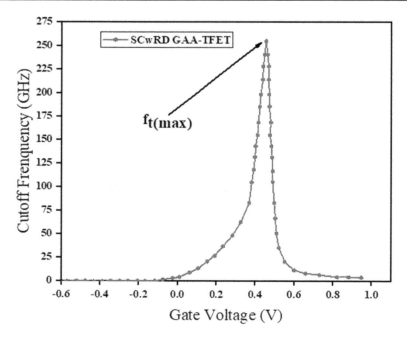

Figure 2.8 SCwRD GAATFET cutoff frequency (f_t) vs gate voltage (Vgs).

Figure 2.8 depicts the cutoff frequency for SCwRD GAA-TFET. A higher cutoff frequency guarantees that a device's analog/RF performance can be enhanced. The stated cutoff frequency for SCwRD GAA-TFET is 260 GHz. Better transconductance and smaller Cgd influence the cutoff frequency of SCwRD GAA-TFET. The cutoff frequency is also known as the gain-band-width product (G_{BW}) or the transition frequency. The greater the cutoff frequency, the superior the device's frequency responsiveness. The higher cutoff frequency also guarantees that the analogue circuit has a wide range for amplification and oscillation. As a result, the device gain will be large for devices with a higher cutoff frequency.

2.5 CONCLUSION

We investigated the architectures of SCw RD GAA-TFET in this study. The device DC and transient efficiency parameters were investigated using the TCAD Synopsys tool. SCwRD's suggested GAA-TFET trumps. The drain-side channel's doping concentration is 100 cm^{-3} higher than the source-side channel's, resulting in improved off-current efficacy. The low I_{OFF} has been reduced to 4.12×10^{-18} A/um.

ACKNOWLEDGMENT

The authors would like to acknowledge the SERB TARE GRANT Project no. TAR/2022/000406, Govt. of India and VIT Bhopal University, Kothrikalan, Sehore-466114 for technical and financial support.

REFERENCES

[1] E. Mollick, "Establishing Moore's law," *IEEE Annals of the History of Computing*, vol. 28, no. 3, pp. 62–75, 2006.

[2] A. C. Seabaugh and Q. Zhang, "Low-voltage tunnel transistors for beyond CMOS logic," *Proceedings of the IEEE*, vol. 98, no. 12, pp. 2095–2110, 2010.

[3] A. Dutt, S. Tiwari, A. Upadhyay, R. Mathew and A. Beohar, "Impact of drain underlap and high bandgap strip on cylindrical gate all around tunnel FET and its influence on analog/RF performance," *Silicon*, Springer, January 2022. https://doi.org/10.1007/s12633-022-01692

[4] S. Tiwari, A. Dutt, M. Joshi, P. Nigam, R. Mathew and A. Beohar, "An investigation of a suppressed-drain cylindrical gate-all-around retrograde-doped heterospacer steep-density-film tunneling field-effect transistor," *Springer Journal of Computational Electronic*, Springer, no. 5, pp. 1702–1710, 2021.

[5] K. Roy, S. Mukhopadhyay and H. Mahmoodi-Meimand, "Leakage current mechanisms and leakage reduction techniques in deep-submicrometer CMOS circuits," *Proceedings of the IEEE*, vol. 91, no. 2, pp. 305–327, 2003.

[6] M. Schlosser, K. K. Bhuwalka, M. Sauter, T. Zilbauer, T. Sulima and I. Eisele, "Fringing-induced drain current improvement in the tunnel field-effect transistor with high-ka gate dielectrics," *IEEE Transactions on Electron Devices*, vol. 56, no. 1, pp. 100–108, 2009.

[7] D. Esseni, M. Guglielmini, B. Kapidani, T. Rollo, and M. Alioto, "Tunnel FETs for ultralow voltage digital VLSI circuits: Part I—Device–circuit interaction and evaluation at device level," *IEEE Transactions on Very Large Scale Integration (VLSI) Systems*, vol. 22, no. 12, pp. 2488–2498, 2014.

[8] A. Dutt, S. Tiwari, M. Joshi, P. Nigam, R. Mathew and A. Beohar, "Diminish short channel effects on cylindrical GAA hetero-gate dielectric TFET using high density delta," *IETE Journal of Research*, 2022. https://doi.org/10.1080/03772063.2022.2081263

[9] Ankur Beohar, Nandakishor Yadav, Ambika Prasad Shah and Santosh Kumar Vishvakarma, "Analog/RF Characteristics of a 3D Cylindrical underlap GAA-TFET based on Ge-Source using fringing field engineering for low power applications," *Springer Journal of Computational Electronics*, vol. 17, no. 4, pp. 1650–1657, 2018.

[10] C. Le Royer and F. Mayer, "Exhaustive experimental study of tunnel field effect transistors (TFETs): From materials to architecture," *10th International Conference on Ultimate Integration of Silicon, ULIS 2009*, 2009, pp. 53–56.

[11] A. Beohar and S. K. Vishvakarma, "Performance enhancement of asymmetrical underlap 3D-cylindrical GAA-TFET with low insertion width," *Micro & Nano Letters*, vol. 11, no. 8, pp. 443–445, 2016.

[12] Ankur Beohar, Nandakishor Yadav, and Santosh Kumar Vishvakarma, "Analysis of Trap Assisted Tunneling in Asymmetrical Underlap 3D-Cylindrical GAA-TFET based on Hetero-Spacer Engineering for Improved Device Reliability," *IET Micro & Nano Letters*, vol. 12, no. 12, pp. 982–986, 2017.

[13] Ambika Prasad Shah, Nandakishor Yadav, Ankur Beohar and Santosh Kumar Vishvakarma, "An efficient NBTI sensor and compensation circuit for stable and reliable SRAM cells," *Elsevier Microelectronics Reliability*, vol. 87, no. 8, pp. 15–23, 2018.

[14] Q. Huang, R. Jia, C. Chen, H. Zhu, L. Guo, J. Wang, J. Wang, C. Wu, R. Wang, W. Bu, et al., "First foundry platform of complementary tunnel-FETs in CMOS baseline technology for ultralow-power IoT applications: Manufacturability, variability and technology roadmap," in *Electron Devices Meeting (IEDM), 2015 IEEE International*, 2015, pp. 22–22.

Chapter 3

Investigation of high-K dielectrics for single- and multi-gate FETs

Sresta Valasa
National Institute of Technology, Warangal, India

Shubham Tayal
Synopsys India Private Ltd., Hyderabad, India

Laxman Raju Thoutam
Amrita Vishwa Vidyapeetham, Kochi, India

3.1 INTRODUCTION

Modern CMOS and integrated circuit (IC) research works are aimed at lowering power dissipations and enhancing the efficiency of single-chip devices [1–3]. In semiconductor technology, the emerging path of very-large-scale integrated (VLSI) devices has played a crucial role. Traditional scaling has dominated semiconductor devices over the past few decades to increase operational speed, lower power consumption, and increase transistor density [4]. Threshold voltage roll-off, higher leakage currents, dopant variations, increased power consumption, degradation in subthreshold swing, and design expenses are some of the drawbacks of the traditional MOSFET [5, 6]. Since the MOSFET acts like a resistor at shorter gate lengths, the drain competes against the gate to take control of the channel. The gate ultimately loses its capability to limit the leakage current pathways that are located at a certain distance from the gate [7, 8]. To address the above issues, and to mitigate the short channel effects, multi-gate MOSFETs such as Double gate, Tri gate, FinFET, Nanowire FET, and Nanosheet FET [9–11] are proposed, which are regarded as promising candidates for the next-generation microelectronic devices.

Furthermore, while concentrating on reduced SCEs, power consumption, and improved efficacy in the device, the channel material employed also needs to be considered. Silicon (Si) has been the dominant channel material for the last few decades due to its profusion in nature, exceptional material qualities with favorable natural oxide silicon dioxide (SiO_2), and decent SiO_2/Si interface [12]. Accordingly, a vast study has been conducted by researchers to examine novel materials and design engineering methods. As a result, high-k materials have been examined to minimize the leakage currents and parasitic capacitances in MOS devices [13–15]. Moreover, the scaling of transistor size

DOI: 10.1201/9781032670270-3

has led to a decrease in SiO_2 gate dielectric thickness to reduce the parasitic capacitance and leakage current. However, when the thickness decreases beyond 5 nm, the leakage current tends to increase significantly, resulting in excessive power consumption and relatively low device efficiency.

A solution to address the aforementioned issues is to supplant the SiO_2 dielectric with a high-k gate dielectric [16, 17]. Several high-k materials, such as Al_2O_3, ZrO_2, HfO_2, Ta_2O_5, and TiO_2, can be used to increase the thickness of gate oxide without a substantial uprise in gate tunneling current [18–20]. Due to the limits imposed by scalability, low-power operations, and unfavorable reactivity with the Si substrate, only a few materials will be preferred for emerging CMOS applications. Enormous breakthroughs have been achieved in the evolution of high-k gate-dielectric materials and metal gate technologies in the past few years [21, 22]. Moreover, when working with high-k materials, it is vital to employ a metal gate with the proper work function. Consequently, this chapter aims to examine the effect of several high-k gate dielectrics on multi-gate FETs.

3.2 IMPORTANCE OF HIGH-K MATERIALS IN CMOS DEVICE SCALING

When fabricating semiconductors, a material with a high-k dielectric constant, one with a dielectric constant greater than SiO_2, may be employed. One of the numerous methods created to enable greater shrinking of microelectronic components, sometimes known as extending Moore's Law, is the use of high-k gate dielectrics [23]. Lower SiO_2 gate dielectric thickness is required to maintain the trend of transistor downscaling. Tunneling-related leakage currents sharply increase as thickness scales drop below 2–3 nm, resulting in excessive power consumption and decreased device performance. Increased gate capacitance can be achieved by switching the SiO_2 gate dielectric for a high-k material [24, 25] as shown in Figure 3.1. Hence, a high-k dielectric material is needed to further continue the scaling for advanced MOS structures. The high-k materials need to possess a number of cutting-edge characteristics along with their high-k value in order to be effective SiO_2 alternatives [26].

a. The gate electrode and Si substrate should be compatible with them chemically.
b. They should possess acceptable interface qualities with the Si substrate and have thermal stability (minimum of 500 °C).
c. The structure may feature a large band gap, a low oxide interface trap density, greater channel mobility, and high band offset energies.
d. Minimal leakage current density and high- dielectric constant ranging between 10 and 50.

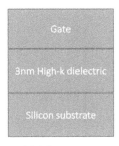 

High-k process
Capacitance =1.6×
Leakage current=0.01 ×

Existing 90nm technology
Capacitance =1×
Leakage current=0.01 ×

Figure 3.1 Comparison of high-k dielectric and conventional gate oxide.

The physical relationship for the thickness of SiO_2 and high-k gate oxide [26] created by the same gate capacitance value (C) is given as

$$C = \frac{\varepsilon_{high-k}}{T_{high-k}} = \frac{\varepsilon_{SiO_2}}{EOT}$$

whereas the thickness of EOT can be given as

$$T_{EOT} = \frac{\varepsilon_{SiO_2}}{\varepsilon_{high-k}} \times T_{phy}$$

A tiny energy band gap results in a low barrier height during the tunneling process in MOSFETs. Therefore, it is crucial to understand the band offsets between a particular high-k insulator and the semiconductor substrate before choosing it for a semiconductor device. It is recommended that the band offset value for high-k dielectrics should be larger than 10, preferably between 25 and 30 [27]. High-k value and band offset do, however, trade off. Figure 3.2 illustrates that the majority of high-k materials have narrower band gaps than SiO_2. For high-K dielectrics, Table 3.1 [24–27] demonstrates the nearly inverse relationship between band gap and dielectric constant. Band gap and dielectric constant are important considerations, but another important element in choosing a high-k material to reach the desired dielectric constant is the oxygen atoms' vibrational frequency. The main issue in microelectronics applications is the minimal vibrational frequency with oxygen bonds. The use of high-k materials in CMOS technology is constrained by the trade-off amid band gap and dielectric constant. Depending on the EOT requirements for a specific technological node, the lower range of permissible dielectric constants may vary.

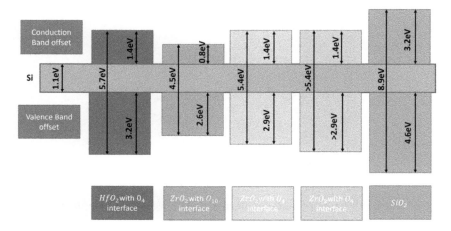

Figure 3.2 Comparison of band gaps and band offsets.

Table 3.1 List of emerging High-k dielectric materials [24–27]

Material	Dielectric value	Crystal structure	Stability with Si	CB offset	Bandgap energy	VB offset
Silicon dioxide (SiO$_2$)	3.9	amorphous	Yes	3.5	8.9	4.4
Aluminum oxide (Al$_2$O$_3$)	9	amorphous	Yes	2.8	8.7	4.9
Yattrium oxide (Y$_2$O$_3$)	15	Cubic	Yes	2.3	5.6	2.6
Tantalum pentoxide (Ta$_2$O$_5$)	26	Orthorhombic	No	0.3	4.4	3.1
Lanthanum oxide (La$_2$O$_3$)	30	Hexagonal, cubic	Yes	2.3	6	0.9
Silicon nitride (Si$_3$N$_4$)	7	Amorphous	Yes	2.4	5.1	1.8
Gadolinium oxide (Gd$_2$O$_3$)	12	Amorphous	Yes	3.2	5.4	3.9
Zirconium dioxide (ZrO$_2$)	25	Monoclinic, cubic, tetragonal	Yes	1.4	7.8	3.3
Hafnium dioxide (HfO$_2$)	25	Monoclinic, cubic, tetragonal	Yes	1.5	5.7	3.4
Titanum dioxide (TiO$_2$)	80	Tetragonal	Yes	1.2	3.5	1.2

Films such as HfO$_2$, ZrO$_2$, and TiO$_2$ appear to be the focus of study right now. ZrO$_2$ and HfO$_2$ were discovered to have leakage currents that were 4–5 orders of magnitude lower than SiO$_2$ with an equivalent gate oxide thickness. Due to similar fabrication chemistry and material characteristics, HfO$_2$ films are said to behave relatively similarly to ZrO$_2$ films (Table 3.2).

Table 3.2 Comparison of various properties of high-k materials [24–27]

Properties	SiO_2	ZrO_2	HfO_2	TiO_2
Structure	Amorphous	Non-crystalline	Non-crystalline	Amorphous, rutile
Silicide formation	NA	Yes	Yes	NA
Formation temperature (°C)	>700	350	350	400
Oxide trap density (Cm^2)	10^{11}	10^{12}	10^{12}	10^{12}
Thermal stability	1000	900	950	550
Interface trap density (Ev-$1Cm^2$)	10^{10}	10^{12}	10^{12}	10^{12}
Low leakage current density w.r.t SiO2	–	10^4–10^5	10^4–10^5	10^1–10^2
Breakdown field (MV/Cm)	10	<4	<4	<4

3.3 EFFECT OF HIGH-K MATERIALS ON MULTI-GATE FETS

To explore the effect of high-k dielectrics on multi-gate FET, junctionless (JL) Nanosheet (NS) FET has been designed to observe the performance. Nanosheet FET is a gate-all-around FET, in which the gate wraps the channel area from all sides, in contrast to that of double-gate FET and FinFET [28]. In order to ease the fabrication process and achieve better performance, a JL structure has been designed here in which the device is doped uniformly avoiding the formation of abrupt junctions [29].

All the simulations necessary for the analysis have been achieved using the industry standard sentaurus TCAD tool [31]. The 3D configuration of the JL-NSFET is visualized in Figure 3.3. In order to validate the accuracy of the results, the device has been calibrated with the experimental data as shown in Figure 3.4 [30]. The simulated data accurately matches the experimental data. Further, several models that include the fermi Dirac model, MLDA model for quantum effects, bandgap narrowing model, SRH, auger recombination models, and Lombardi and Philips unified mobility models [32] have been invoked to assess the device performance.

3.3.1 Effect of high-k materials for various temperatures

This subsection effectively addresses the performance of high-k materials at different temperatures. Here gate stack technology (GS) is used wherein a layer of TiO_2 (k = 40) is deposited on SiO_2 (interfacial layer). The temperature plays a crucial role in the design of a device, and the device must withstand a certain range of temperatures according to the type of intended

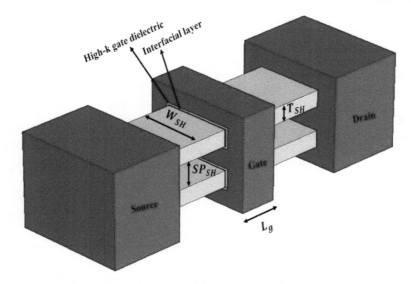

Figure 3.3 Three-dimensional view of JL-NSFET.

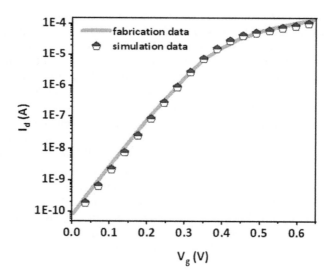

Figure 3.4 Calibration of experimental data [30].

application such as aerospace, nuclear industry, defense, satcoms, telecommunication networks, IR sensors, etc. [33]. To analyze the device performance, the gate length, thickness, and width of the JL-NSFET have been set to 10 nm, 6 nm, and 18 nm, respectively, according to IRDS [34]. The gate workfunction, doping concentration, and EOT have been fixed to 4.8 eV,

1×10^{19} cm^{-3}, and 0.7 nm, respectively. Figure 3.5 portrays the transfer and analog/RF characteristics of JL-NSFET for temperatures of 100°K and 400°K. Figure 3.5(a) illustrates that TiO$_2$ gives improved drain current (I$_d$) in comparison to SiO$_2$ for given temperatures. However, degradation

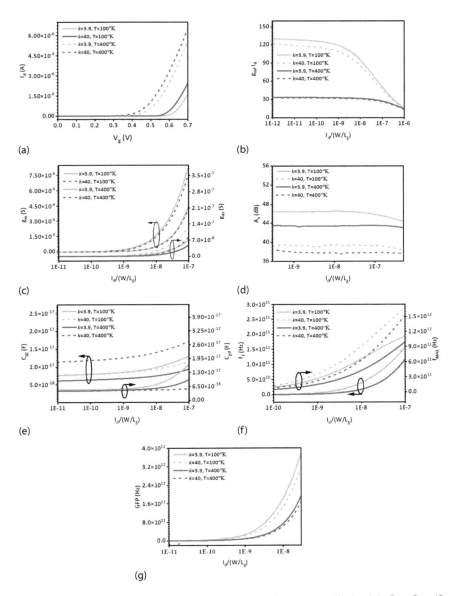

Figure 3.5 (a) Transfer characteristics; (b) g_m/I_d; (c) g_m, g_{ds}; (d) A_v; (e) C_{gg}, C_{gd}; (f) f_T, f_{MAX}; (g) GFP.

in transconductance efficiency (g_m/I_d) has been noticed for HfO_2 material (Figure 3.5(b)) due to the relative degradation in transconductance (g_m) (Figure 3.5(c)) as compared to the improvement in I_d.

It can also be seen that employing TiO_2 as high-k is degrading the output conductance (g_{ds}) for given temperatures, which further adversely impacts the intrinsic gain (Av) by deteriorating it (Figure 3.5(d)). The gate capacitance (C_{gg}) and gate-to-drain capacitance (C_{gd}) are the two prominent factors that determine any device RF functionality. From Figure 3.5(e), it is seen that TiO_2 material is reducing the capacitances, which is in fact a good sign. Nevertheless, the lowering in C_{gg} is proportionate to the degradation in gm, and hence gives a trivial change in cutoff frequency (f_T) as portrayed in Figure 3.5(f). However, the improvement in C_{gd} is fetching a better maximum oscillation frequency (f_{MAX}) for HfO_2 as compared to SiO_2. Moreover, Figure 3.5(g) shows that TiO_2 materials degrade the gain frequency product (GFP) for given temperatures as compared to SiO_2 owing to the heavy deterioration in gm in comparison with f_T. Consequently, it is suggested that high-k material in GS configuration at lower temperatures (100°K) is suitable for RF applications whereas SiO_2 can be preferred for analog applications from the above analysis.

3.3.2 Effect of high-k dielectric spacer materials

To minimize the charge sharing [35, 36] that occurs in the device, spacers are necessary to achieve gate controllability of the device. However, the use of traditional SiO_2 as the spacer material increases the series resistance and reduces the ON current of the device. Considering this, high-k spacers are adopted to enhance the ON current. In this subsection we are going to study the performance of several high-k spacer materials possessing different dielectric constants. Starting with conventional SiO_2, materials such as Si_3N_4, HfO_2, and TiO_2, having dielectric constants k = 3.9, 7.5, 22, and 40, respectively are examined. In order to target the lower gate lengths, here L_g is fixed to 5 nm whereas the T_{SH} and W_{SH} are fixed at 4 nm and 18 nm, respectively.

The spacer length (L_{sp}) has been taken as 18 nm. Figure 3.6 shows the transfer and analog/RF characteristics of several spacer materials relating to normalized drain current ($I_d/(W/L_g)$). It can be observed that the usage of high-k materials lowers the I_{OFF} (Figure 3.6(a)) owed to the larger vertical electric field in the OFF-state region in conjunction with the higher band energies (higher barrier height) (Figure 3.6(c)) along the channel region, which would further enhance the electrostatic integrity of the gate. The employment of TiO_2 material gives enhanced g_m/I_d (Figure 3.6(b)) as compared to other three materials. From Figure 3.6(d) it is witnessed that g_m and g_{ds} are low for TiO_2 material. This improvement in g_{ds} implies the reduction in channel control by drain, and this consecutively minimizes the channel length modulation (CLM) effect [37]. Moreover, this improvement in g_{ds} will further enhance the A_V for TiO_2 material (Figure 3.6(e)). The high 'k'

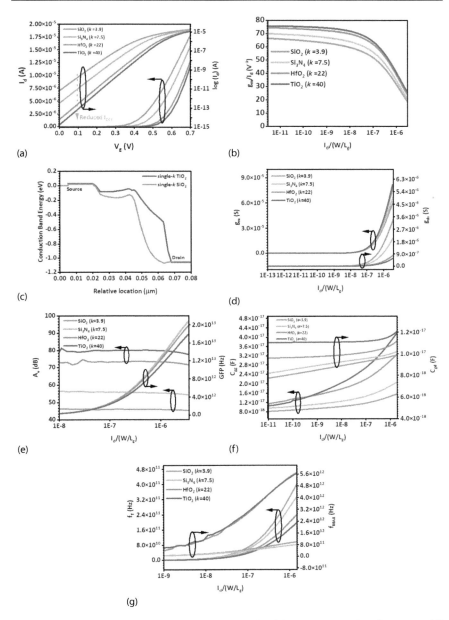

Figure 3.6 (a) Transfer characteristics; (b) g_m/I_d; (c) conduction band energy; (d) g_m, g_{ds}; (e) A_v, GFP; (f) C_{gg}, C_{gd}; (g) f_T, f_{MAX}.

value of the material escalates the fringing fields of the JLNSFET, which in turn upsurges C_{gg} and C_{gd}. This degradation in C_{gg} will reduce the f_T for TiO_2 material whereas the lower g_{ds} values dominates the degradation in C_{gd} and hence improves the f_{MAX} of the device (Figure 3.6(f)). Furthermore, GFP is noticed to be degraded for TiO_2 material owing to the degradation in f_T as shown in Figure 3.6(e). Consequently, it is suggested to use high-k material for analog applications whereas low-k material for RF operations for JL-NSFET.

3.4 EFFECT OF DUAL-K SPACER DIELECTRIC MATERIALS

The main issue with the high-k spacer dielectric material is the delay in the cirucits caused due to fringe-related capacitances. Moreover, the formation of traps degrades the carrier mobility of the device. One effective way to solve this is the utilization of dual dielectric spacer material having the combination of inner high dielectric and outer low dielectric value [38, 39]. For the purpose of investigation, four combinations are taken into consideration having inner high-k + outer low-k and inner low-k + outer high-k. HfO_2 and TiO_2 having dielectric constant k = 22 and 40, respectively, are employed as high-k dielectric spacers, and Si_3N_4 (k = 7.5) is employed as low-k dielectric spacer.

It can be noticed from Figure 3.7(a) that inner high-k + outer low-k combination reduces the OFF current and increases the drive current as compared to all other combinations and single high-k dielectric spacer materials. This improvement in drive current is experienced because the high-k dielectric material incorporated near the channel region initiates the fringing fields to increase the carrier density. The decrease in OFF current is due to the increase in height of the barrier (Figure 3.7(b)). The g_m/I_d is noticed to be improved for k = 40 + 7.5, ensuring enhanced input drivability of the device as shown in Figure 3.7(c). The improvement in gm and gds for k = 40 + 7.5 (Figure 3.7(d)) further improves the A_V of the device for k = 40 + 7.5 as portrayed in Figure 3.7(e). C_{gg} and C_{gd} are noticed to be lowered with inner high-k + outer low-k combination as depicted in Figure 3.7(f). This reduction in C_{gg} and C_{gd} further enhances the f_T and f_{MAX} of the device for inner high-k + outer low-k combination as illustrated in Figure 3.7(g). Consequently, it is suggested that inner high-k + outer low-k offers enhanced performance in comparison to single high-k dielectric spacer materials. k = 40 + 7.5 can be chosen for analog applications, and k = 22 + 7.5 as well as k = 40 + 7.5 can be chosen for RF applications.

3.5 CONCLUSION

This work reports on the high-k dielectric materials for multi-gate FET devices. Junctionless NSFET has been reported with gate stack technology and spacer technology. As far as gate stack technology is concerned, HfO_2 (k = 22) material at lower temperatures (100°K) is suitable for RF operations

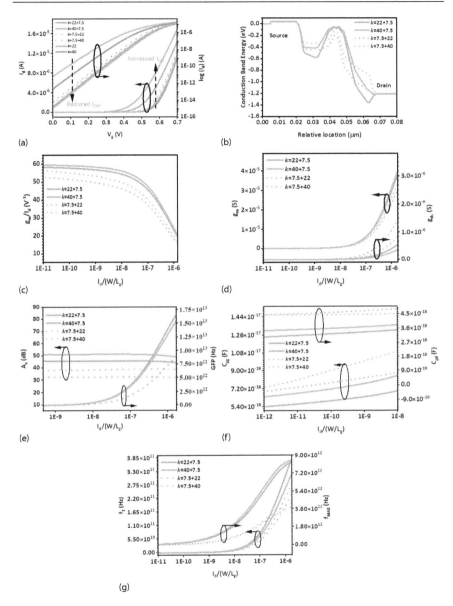

Figure 3.7 (a) Transfer characteristics; (b) conduction band energy; (c) g_m/I_d; (d) g_m, g_{ds}; (e) A_v, GFP; (f) C_{gg}, C_{gd}; (g) f_T, f_{MAX}.

whereas SiO_2 can be utilized for analog applications. The use of traditional SiO_2 as the spacer dielectric material increases the series resistance and reduces the ON current of the device, and hence high-k dielectric spacers are adopted to improve the drive current. The analysis revealed high-k spacer materials can be preferred for analog applications whereas low-k material for RF operations for JL-NSFET. Dual dielectric spacer materials are also investigated, and it is suggested to adopt k = 40 + 7.5 for analog applications and k = 22 + 7.5 as well as k = 40 + 7.5 for RF applications.

REFERENCES

[1] Y. Taur et al., "CMOS scaling into the nanometer regime," *Proc. IEEE*, vol. 85, no. 4, pp. 486–503, 1997, doi:10.1109/5.573737

[2] M. Horowitz, E. Alon, D. Patil, S. Naffziger, R. Kumar, and K. Bernstein, "Scaling, power, and the future of CMOS," *Tech. Dig. - Int. Electron Devices Meet. IEDM*, vol. 2005, pp. 9–15, 2005, doi:10.1109/vlsid.2007.140

[3] S. Valasa, J. R. Shinde, D. R. Ramji, and S. Avunoori, "A Power and Delay Efficient Circuit for CMOS Phase Detector and Phase Frequency Detector," *Proc. 6th Int. Conf. Commun. Electron. Syst. ICCES 2021*, pp. 77–82, 2021, doi:10.1109/ICCES51350.2021.9489140

[4] R. Gonzalez, B. M. Gordon, and M. A. Horowitz, "Supply and threshold voltage scaling for low power CMOS," *IEEE J. Solid-State Circuits*, vol. 32, no. 8, pp. 1210–1216, 1997, doi:10.1109/4.604077

[5] T. Endoh and Y. Norifusa, "Scalability of vertical MOSFETs in sub-10nm generation and its mechanism," *IEICE Trans. Electron.*, vol. E92-C, no. 5, pp. 594–597, 2009, doi:10.1587/transele.E92.C.594

[6] E. Rauly, B. Iniguez, D. Flandre, and C. Raynaud, "Investigation of single and double gate SOI MOSFETs in accumulation Mode for enhanced performances and reduced technological drawbacks," *Eur. Solid-State Device Res. Conf.*, pp. 540–543, 2000, doi:10.1109/ESSDERC.2000.194834

[7] A. Akturk et al., "Compact and distributed modeling of cryogenic bulk MOSFET operation," *IEEE Trans. Electron Devices*, vol. 57, no. 6, pp. 1334–1342, 2010, doi:10.1109/TED.2010.2046458

[8] A. Khakifirooz, O. M. Nayfeh, and D. Antoniadis, "A simple semiempirical shortchannel MOSFET current-voltage model continuous across all regions of operation and employing only physical parameters," *IEEE Trans. Electron Devices*, vol. 56, no. 8, pp. 1674–1680, 2009, doi:10.1109/TED.2009.2024022

[9] D. Bhattacharya and N. K. Jha, "FinFETs: From devices to architectures," *Adv. Electron.*, vol. 2014, pp. 1–21, 2014, doi:10.1155/2014/365689

[10] N. Guduri, D. Kannuri, R. R. Maram, N. K. Kevuloth, S. Valasa, and S. Tayal, "Performance Analysis of Dielectrically Separated Independent Gates Junctionless DG-MOSFET: A Digital Perspective," *2022 IEEE Int. Conf. Nanoelectron. Nanophotonics, Nanomater. Nanobioscience Nanotechnology, 5NANO 2022*, 2022, doi:10.1109/5NANO53044.2022.9828882

[11] J. Ajayan et al., "Nanosheet field effect transistors-A next generation device to keep Moore's law alive: An intensive study," *Microelectronics J.*, vol. 114, 2021, doi:10.1016/j.mejo.2021.105141

[12] S. Valasa, S. Tayal, L. R. Thoutam, J. Ajayan, and S. Bhattacharya, "A critical review on performance, reliability, and fabrication challenges in nanosheet FET for future analog/digital IC applications," *Micro Nanostructures*, vol. 170, p. 207374, 2022, doi:10.1016/j.micrna.2022.207374

[13] M. Houssa, "High k Gate Dielectrics - Google Livres," 2003, [Online]. Available: https://books.google.co.ma/books?hl=fr&lr=&id=DibOBgAAQBAJ&oi= fnd&pg=PP1&dq=Houssa+M+2003+High+k+Gate+Dielectrics+(CRC+ Press)&ots=ex_jLEvTkl&sig=gnSGhg3yVI3h6uBaGUEx-km1O6U&redir_ esc=y#v=onepage&q&f=false

[14] R. M. C. De Almeida and I. J. R. Baumvol, "Reaction-diffusion in high-k dielectrics on Si," *Surf. Sci. Rep.*, vol. 49, no. 1–3, pp. 1–114, 2003, doi:10.1016/S01675729(02)00113-9

[15] G. He, X. Chen, and Z. Sun, "Interface engineering and chemistry of Hf-based high-k dielectrics on III-V substrates," *Surf. Sci. Rep.*, vol. 68, no. 1, pp. 68–107, 2013, doi:10.1016/j.surfrep.2013.01.002

[16] V. Goyal, S. Tayal, S. Meena, R. Gupta, and S. Bhattacharya, "High-k gate dielectrics and metal gate stack technology for advance semiconductor devices," *High-k Mater. Multi-Gate FET Dev.*, pp. 19–32, 2021, doi:10.1201/9781003121589-2

[17] S. Hall, O. Buiu, I. Z. Mitrovic, Y. Lu, and W. M. Davey, "Review and perspective of high-k dielectrics on silicon," *J. Telecommun. Inf. Technol.*, vol. 2, pp. 33–43, 2007.

[18] O. Engström et al., "Navigation aids in the search for future high-k dielectrics: Physical and electrical trends," *Solid. State. Electron.*, vol. 51, no. 4, pp. 622–626, 2007, doi:10.1016/j.sse.2007.02.021

[19] S. Valasa, S. Tayal, and L. R. Thoutam, "Design insights into thermal performance of vertically stacked JL-NSFET with high-k gate dielectric for Sub 5-nm technology node," *ECS J. Solid State Sci. Technol.*, vol. 11, no. 4, p. 041008, 2022, doi:10.1149/2162-8777/ac6627

[20] G. He and Z. Sun, "High-k Gate Dielectrics for CMOS Technology," *High-k Gate Dielectr. C. Technol.*, 2012, doi:10.1002/9783527646340

[21] F. Palumbo et al., "A review on dielectric breakdown in thin dielectrics: Silicon dioxide, high-k, and layered dielectrics," *Adv. Funct. Mater.*, vol. 30, no. 18, 2020, doi:10.1002/adfm.201900657

[22] V. Narendar, "Performance enhancement of FinFET devices with gate-stack (GS) high-K dielectrics for nanoscale applications," *Silicon*, vol. 10, no. 6, pp. 2419–2429, 2018, doi:10.1007/s12633-018-9774-7

[23] M. K. Bera and C. K. Maiti, "Electrical properties of SiO2/TiO2 high-k gate dielectric stack," *Mater. Sci. Semicond. Process.*, vol. 9, no. 6, pp. 909–917, 2006, doi:10.1016/j.mssp.2006.10.008

[24] G. Jun and K. Cho, "First principles modeling of high-K dielectric materials," *Mater. Res. Soc. Symp. - Proc.*, vol. 747, pp. 83–91, 2003, doi:10.1557/proc-747-t5.1/n7.1

[25] R. Chau, "Advanced Metal Gate/High-K Dielectric Stacks for High-Performance CMOS Transistors," *AVS 5th Int. Microelectron. Interfaces Conf*, vol. 90, pp. 9–11, 2004, [Online]. Available: http://www.intel.com/content/dam/doc/white-paper/high-kgate-dielectrics-for-cmos-transistors-paper.pdf

[26] H. Wong and H. Iwai, "On the scaling issues and high-κ replacement of ultra-thin gate dielectrics for nanoscale MOS transistors," *Microelectron. Eng.*, vol. 83, no. 10, pp. 1867–1904, 2006, doi:10.1016/j.mee.2006.01.271

[27] P. Chowdhury et al., "The structural and electrical properties of TiO_2 thin films prepared by thermal oxidation," *Phys. B Condens. Matter*, vol. 403, no. 19–20, pp. 3718–3723, 2008, doi:10.1016/j.physb.2008.06.022

[28] S. Valasa, S. Tayal, and L. R. Thoutam, "An intensive study of tree-shaped JLNSFET: Digital and analog/RF perspective," *IEEE Trans. Electron Devices*, pp. 1–8, 2022, doi:10.1109/ted.2022.3216821

[29] S. Valasa et al., "Design and performance optimization of junctionless bottom spacer FinFET for digital/analog/RF applications at Sub-5nm technology node," *ECS J. Solid State Sci. Technol.*, doi:10.1149/2162-8777/acb175

[30] N. Loubet et al., "Stacked nanosheet gate-all-around transistor to enable scaling beyond FinFET," *Dig. Tech. Pap. - Symp. VLSI Technol.*, pp. T230–T231, 2017, doi: 10.23919/VLSIT.2017.7998183

[31] https://www.synopsys.com/silicon/tcad/device-simulation/sentaurus-device.html

[32] S. Tayal et al., "Investigation of nanosheet-FET based logic gates at Sub-7 nm technology node for digital IC applications," *Silicon*, 2022, doi:10.1007/s12633022-01934-x

[33] V. B. Sreenivasulu and V. Narendar, "Design and temperature assessment of junctionless nanosheet FET for nanoscale applications," *Silicon*, vol. 14, no. 8, pp. 3823–3834, 2022, doi:10.1007/s12633-021-01145-w

[34] https://irds.ieee.org/editions/2021/systems-and-architectures

[35] Y. C. Eng, J. T. Lin, K. D. Huang, T. Y. Lee, and K. C. Lin, "An investigation of the effects of Si thickness-induced variation of the electrical characteristics in FDSOI with block oxide," *ICSICT-2006 2006 8th Int. Conf. Solid-State Integr. Circuit Technol. Proc.*, pp. 61–64, 2006, doi:10.1109/ICSICT.2006.306077

[36] S. Valasa, S. Tayal, and L. R. Thoutam, "Performance evaluation of spacer dielectric engineered vertically stacked junctionless nanosheet FET for Sub-5 nm technology node". *ECS J. Solid State Sci. Technol.*, vol. 11, no. 9, p. 093006, 2022, doi:10.1149/2162-8777/ac90ec

[37] B. Ghosh, P. Mondal, M. W. Akram, P. Bal, and A. K. Salimath, "Heterogatedielectric double gate junctionless transistor (HGJLT) with reduced band-to-band tunnelling effects in subthreshold regime," *J. Semicond.*, vol. 35, no. 6, 2014, doi:10.1088/1674-4926/35/6/064001

[38] P. K. Pal, B. K. Kaushik, and S. Dasgupta, "Asymmetric dual-spacer trigate FinFET device-circuit codesign and its variability analysis," *IEEE Trans. Electron Devices*, vol. 62, no. 4, pp. 1105–1112, 2015, doi:10.1109/TED.2015.2400053

[39] P. K. Pal, B. K. Kaushik, and S. Dasgupta, "Investigation of symmetric dual-k spacer trigate FinFETs from delay perspective," *IEEE Trans. Electron Devices*, vol. 61, no. 11, pp. 3579–3585, 2014, doi:10.1109/TED.2014.2351616

Chapter 4

Measurement of back-gate biasing for ultra-low-power subthreshold logic in FinFET device

Ajay Kumar Dadoria
Amity University, Gwalior, India

Uday Panwar and Narendra Kumar Garg
SIRT Bhopal, Bhopal, India

4.1 INTRODUCTION

Nowadays, with the goal of minimizing power consumption, the size of routing interconnection in electronic devices such as cellular phones, multimedia devices, and personal notebooks has been compacted as more and more functionality is incorporated into these devices. This is necessary because these devices rely on battery life. However, as devices are scaled down, the overall chip area is reduced, and more transistors are incorporated onto a single chip die area to maintain Moore's law of scaling. This results in increased power consumption due to the fabrication of millions of transistors on a single IC. Power consumption is determined by the technology node used, with above 90 nm being dynamic power dominant and below 65 nm being static or leakage power dominant with technology scaling. Two types of scaling are considered to maintain the performance of the circuit, namely constant voltage scaling and constant field scaling. Dynamic power is directly proportional to the supply voltage, so reducing power consumption significantly improves the overall performance of the circuit. Although scaling reduces transistor count, there is an increase in the evaluation delay of the circuit, which reduces the clock frequency [1]. To improve the performance of circuits, it is necessary to introduce new techniques that can mitigate power consumption and enhance speed [2]. One way to achieve this is by using a multi-VDD system, where the critical path is provided with a standard power supply (VDDH), while the noncritical path is scaled down to a lower supply voltage (VDDL) [3, 4]. This helps maintain the supply voltage of the critical path and overcome speed constraints in the design. Further, an architecture for high-performance low-power wide fan-in dynamic or gates was analyzed [5, 6]. By reducing stand-by power and evaluation delay, the overall performance of the circuit can be improved.

DOI: 10.1201/9781032670270-4

49

In Section 4.2, we will take about some of the properties of FinFET. The results and discussion are shown in Section 4.3. Finally, conclusions are presented in Section 4.4.

4.2 FINFET TECHNOLOGY

A new device has been developed to address the problem of shorter channel effect and reduce leakage power. This device is called a multigate field effect transistor (FET), also known as the FinFET, which utilizes vertical channel constructions resembling fish fins. Compared to bulk transistors, the FinFET's fins are usually lightly doped or undoped, resulting in improved carrier mobility and doping fluctuation. Figure 4.1(a) shows the general plane cross-section view of a double-gate FinFET (DGFET), while Figure 4.1(b) displays the overlapped gate fin. FinFET devices offer better performance and lower power consumption, resulting in a larger I_{ON}/I_{OFF} ratio, decreased leakage current, and faster circuit switching speed.

Figure 4.1(a) depicts the three-dimensional architecture of the FinFET. The device is referred to as quasi-planar in FinFET because the channel is constructed perpendicular to the wafer plane and current flows parallel to the wafer plane [3]. The FinFET's effective channel width is denoted by the symbol Wmin and given by the equation:

$$W_{min} = 2H_{fin} + T_{fin}$$

Effective channel length is $L_{eff} = L_{gate} + 2 \times L_{ext}$
where (Hfin > tsi), the height of the fin is always more than the thickness of the fin, suppressing the effects of shorter channels and reducing the area of DGFET [11]. When there are n parallel fins, the overall width is denoted by

$$W_{Total} = n\left(2H_{fin} + t_{si}\right)$$

FinFET better optimizes the subthreshold leakage current and improves the performance to mitigate power consumption. FinFET can be categorized

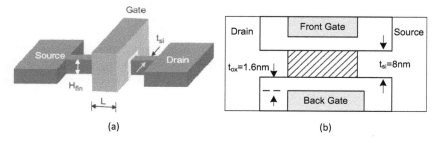

(a)   (b)

Figure 4.1 (a) FinFET 3D view of single fin. (b) Top view of FinFET.

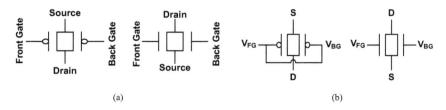

Figure 4.2 (a) DGPMOS and DGNMOS FinFET structure. (b) SG mode (V_{FG} = V_{BG}), IG mode ($V_{FG} \neq V_{BG}$) configuration of FinFET.

Table 4.1 Performance parameters of FinFET device with different technology

Parameter	Technology				
	7 nm	*10 nm*	*14 nm*	*16 nm*	*20 nm*
H_{FIN} (nm)	18	22	24	27	30
W_{FIN} (nm)	12	15	16	22	26
L (nm)	8	12	16	18	20
V_{dd}	0.8 V	0.78 V	0.9 V	0.95 V	0.8 V

in two symmetric double gates, which have three terminal sources, drain, and gate, as shown in Figure 4.2(a). When both front and back gates are tied together ((V_{FG} = V_{BG} = V_{dd}), we apply the same voltage to both; the gate is known as SG mode or symmetric double gate (3T), as shown in Figure 4.2(b). In independent gate mode or asymmetric double gate, FG and BG are at different potential ($V_{FG} \neq V_{BG}$) and both gates are asymmetric, as shown in Figure 4.2(b); both gates are connected independently with various configurations [7, 8]. Asymmetric dual gate has excellent control over the channel, which reduces I_{OFF} current, increases I_{ON} current, achieves ideal subthreshold slope, etc. Asymmetric DGMOS with independent biasing of FG and BG helps lower dynamic threshold voltage, which helps lowe I_{OFF} and increase I_{ON} current with low gate capacitance, to achieve high design flexibility at the circuit level [9, 10]. Table 4.1 shows different parameters to construct the model of FinFET from 7-nm to 20-nm technology transistor with variation of supply voltage with scaling of technology and 7T SRAM cell with high write and read margins for sub-20 nm FinFET technologies [11–15].

4.3 RESULTS AND DISCUSSION

This study involved measuring the I$_{ON}$ and I$_{OFF}$ currents of a FinFET device while keeping the front gate voltage constant and varying the back gate. Simulations were conducted using the FinFET technology in HP and LSTP models at different technology nodes (20 nm, 16 nm, 14 nm, 10 nm, and 7 nm) to mitigate power dissipation in the I$_{ON}$ and I$_{OFF}$ currents. HSPICE was used for simulations, with all parameters varied in DSM technology.

From Table 4.1, we calculated the I$_{ON}$ and I$_{OFF}$ currents for N-FinFET and P-FinFET from 20-nm to 7-nm technology. Simulation results showed that increasing the reverse voltage or back gate bias from −0.2 V to −0.4 V had a negligible impact on the variation of I$_{ON}$ and I$_{OFF}$ current. However, as technology was reduced from 20 nm to 7 nm, there was a slight reduction in the I$_{ON}$ current from 1.712 to 1.559 μA. The reduction was small compared to the reduction in technology, but there was a significant decrease in the value of I$_{OFF}$ current from 36.28 to 23.96 nA. Reducing the I$_{OFF}$ current results in improving the driving capability of the transistor. Table 4.2 shows that the impact on I$_{ON}$ current is similar, with a reduction from 1.591 to 1 μA. This reduction is very small with the reduction in technology, but there is a drastic reduction in the value of I$_{OFF}$ current, which decreases from 32.65 to 21.17 nA when we reduce the I$_{OFF}$ current in the HP model of FinFET technology. Table 4.3 shows an increase in the I$_{ON}$ current from 1.702 to 526.6 μA, which improves the driving capability of digital ICs, but I$_{OFF}$ current also reduces from microampere to picoampere (34.67–23.87 pA). However, as we move from 20 nm to 7 nm, there is less reduction in the I$_{OFF}$ current in the LSTP model for N-FinFET. Table 4.4 shows the same trend for P-FinFET, where I$_{ON}$ current increases from 455.7 to 725.0 μA in the LSTP model. From Tables 4.1 to 4.4 there is no variation in I$_{ON}$ and I$_{OFF}$ current of N-FinFET and P-FinFET in both HP and LSTP models of

Table 4.2 Calculation of I$_{ON}$ and I$_{OFF}$ current in N-FinFET by using HP model

VFG	VFG = VBG		VBG = 0		VBG = −0.2		VBG = −0.4	
	I$_{ON}$ (μA)	I$_{OFF}$ (nA)	I$_{ON}$ (μA)	I$_{OFF}$ (nA)	I$_{ON}$ (μA)	I$_{OFF}$ (nA)	I$_{ON}$ (μA)	I$_{OFF}$ (nA)
7 nm	1.669	24.96	1.769	24.96	1.669	25.96	1.769	32.96
10 nm	1.731	29.58	1.831	29.58	1.731	25.58	1.831	22.58
14 nm	1.701	32.06	1.701	33.06	1.701	32.06	1.701	32.06
16 nm	1.601	34.90	1.801	34.90	1.701	34.90	1.701	35.90
20 nm	1.802	37.28	1.902	37.28	1.902	37.28	1.602	37.28

Table 4.3 Calculation of I$_{ON}$ and I$_{OFF}$ current in P-FinFET by using HP model

VFG	VFG = VBG		VBG = 0		VBG = −0.2		VBG = −0.4	
	I$_{ON}$ (μA)	I$_{OFF}$ (nA)	I$_{ON}$ (μA)	I$_{OFF}$ (nA)	I$_{ON}$ (μA)	I$_{OFF}$ (nA)	I$_{ON}$ (μA)	I$_{OFF}$ (nA)
7 nm	1.518	23.17	1.518	23.17	1.518	23.17	1.518	23.17
10 nm	1.509	25.72	1.509	25.72	1.509	25.72	1.509	25.72
14 nm	1.458	24.65	1.458	24.65	1.458	24.65	1.458	24.65
16 nm	1.586	32.01	1.586	32.01	1.586	32.01	1.586	32.01
20 nm	1.691	34.65	1.691	34.65	1.691	34.65	1.691	34.65

Table 4.4 Calculation of I_{ON} and I_{OFF} current in N-FinFET by using LSTP model

VFG	VFG = VBG		VBG = 0		VBG = −0.2		VBG = −0.4	
	I_{ON} (μA)	I_{OFF} (pA)	I_{ON} (μA)	I_{OFF} (pA)	I_{ON} (pA)	I_{OFF} (pA)	I_{ON} (μA)	I_{OFF} (pA)
7 nm	536.6	25.87	536.6	25.87	536.6	25.87	536.6	25.87
10 nm	654.5	29.51	654.5	29.51	654.5	29.51	654.5	29.51
14 nm	718.4	32.21	718.4	32.21	718.4	32.21	718.4	32.21
16 nm	771.4	35.53	771.4	35.53	771.4	35.53	771.4	35.53
20 nm	824.3	36.67	824.3	36.67	824.3	32.67	824.3	36.67

Table 4.5 Calculation of I_{ON} and I_{OFF} current in P-FinFET by using LSTP model

VFG	VFG = VBG		VBG = 0		VBG = −0.2		VBG = −0.4	
	I_{ON} (μA)	I_{OFF} (pA)	I_{ON} (μA)	I_{OFF} (pA)	I_{ON} (pA)	I_{OFF} (pA)	I_{ON} (μA)	I_{OFF} (pA)
7 nm	455.7	21.78	455.7	21.78	455.7	21.78	455.7	21.78
10 nm	553.7	24.84	553.7	24.84	553.7	24.84	553.7	24.84
14 nm	622.1	27.34	622.1	27.34	622.1	27.34	622.1	27.34
16 nm	679.5	31.88	679.5	31.88	679.5	31.88	679.5	31.88
20 nm	725.0	32.47	725.0	32.47	725.0	32.47	725.0	32.47

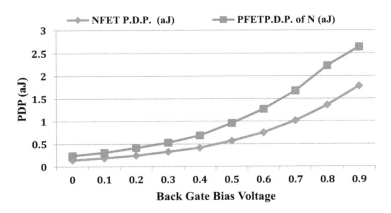

Figure 4.3 Power delay product of back gate biase.

FinFET technology when the back gate is biased. Table 4.5 depicts the analysis of I_{ON} and I_{OFF} current in P-FinFET by using LSTP model

4.4 CONCLUSION

This study focuses on evaluating the electrical characteristics of FinFET circuits by calculating the IOFF and ION currents of both N-FinFET and

P-FinFET in the LSTP and HP models, ranging from 7-nm to 20-nm technology. The effects of back gate bias were also examined, and it was found that there was no impact on IOFF and ION currents in either model. In the LSTP model, the ION current was larger than the HP model due to the use of a lower stand-by power library that suppresses unwanted leakage current. The results suggest that FinFET-based leakage reduction approaches could be useful for future applications that demand low power consumption. Additionally, the ION current of P-FinFET increased from 455.7 to 725.0 µA in the LSTP model. Lastly, the ION and IOFF current of both N-FinFET and P-FinFET in both the HP and LSTP models remained constant when the back gate was biased. It should be noted that all designs included an additional leakage power saving transistor, which increased the circuit's area.

REFERENCES

[1] H.-S. P. Wong, D. J. Frank, P. M. Solomon, C. H.-J. Wann, and J. J. Welser, "Nanoscale CMOS", *Proceedings of IEEE*, Vol. 87, pp. 537–570, 1999.

[2] E. Nowak, et al. "Turning silicon on its edge", *IEEE Circuits & Device Magazine*, pp. 20–31, (2004).

[3] S. A. Tawfika, and V. Kursun, "FinFET domino logic with independent gate keepers", *Micro Electronics Journal*, Vol. 40, pp. 1531–1540 (2009).

[4] A. K. Dadoria, K. Khare, T.K. Gupta, and R.P. Singh, "Leakage reduction by using FinFET technique for nanoscale technology circuits", *Journal of Nanoelectronics and Optoelectronics*, Vol. 12, no. 3, pp. 1–8, 2017.

[5] L. Garg, and V. Sahula, "Macromodels for static virtual ground voltage estimation in power-gated circuits", *IEEE Transactions on Circuits and Systems—II: Express Briefs*, Vol. 63, no. 5, pp. 468–472, 2016.

[6] H. F. Dadgour, and K. Banerjee, "A novel variation-tolerant keeper architecture for high-performance low-power wide fan-in dynamic or gates", *IEEE Transaction Very Large Scale Integration (VLSI) System*, Vol. 18, no. 2, pp. 1567–1577, 2010.

[7] W. Hisamoto, W.C. Lee, J. Kedzierski et al. "FinFET-A self-aligned double-gate MOSFET scalable to 20 nm" *IEEE Transaction on Electron Devices*, Vol. 48, no. 10, pp. 2357–2362, 2001.

[8] P. Mishra, A. Muttreja, and N. K. Jha, "FinFET circuit design, Springer", *Nano-electronic Circuit Design*, pp. 23–53, 2011. doi:10.1007/978-1-4419-7609-3_2

[9] M. W. Allah, M. H. Anis, and M. I. Elmasry, High speed dynamic logic circuits for scaled-Down CMOS and MTCMOS technologies, in *Proceedings of IEEE International Symposium on Low Power Electronics Design*, pp. 123–128 2000.

[10] A. Tawfik, and V. Kursun, "High speed FinFET domino logic circuits with independent gate-biased double-gate keepers providing dynamically adjusted immunity to noise", in *Proceedings of IEEE International Conference on Microelectronics*, pp. 175–178, 2007.

[11] Y. Liu et al. "Cointegration of high-performance tied-gate three-terminal FinFETs and variable threshold-voltage independent-gate four-terminal FinFETs with asymmetric gate-oxide thicknesses", *IEEE Electron Device Letters*, Vol. 28, no. 6, pp. 517–519, 2007.

[12] P. Mishra, and N.K. Jha, "Low-power FinFET circuit synthesis using surface orientation optimization", in Proceedings of *Design Automation and Test in Europe*, pp. 311–314, 2010.

[13] J. J. Kim, and K. Roy, "Double gate-MOSFET sub-threshold circuit for ultralow power applications', *IEEE Transaction on Electron Devices*, Vol. 51, no. 9, pp. 1468–1474, 2004.

[14] A. K. Dadoria, K. Khare, T. K. Gupta, and R. P. Singh, "Ultra-low power FinFET-based domino circuits" *International Journal of Electronics*, Vol. 104, pp. 952–967, 2017.

[15] M. Ansari, H. Afzali-Kusha, B. Ebrahimi, Z. Navabi, A. Afzali-Kusha, and M. Pedram, "A near-threshold 7T SRAM cell with high write and read margins and low write time for sub-20 nm FinFET technologies", *Integration of VLSI*, Vol. 50, pp. 91–106, 2015.

Chapter 5

Compact analytical model for graphene field effect transistor

Drift-diffusion approach

Abhishek Kumar Upadhyay

X-FAB Dresden Grenzstraße 28, Dresden, Germany

Siromani Balmukund Rahi

Gautam Buddha University, Greater Noida, India

Billel Smaani

Centre Universitaire Abdelhafid Boussouf, Mila, Algeria

Ball Mukund Mani Tripathi

Velagapudi Ramakrishna Siddhartha Engineering College, Vijayawada, India

Neha Paras

National Institute of Technology, Delhi, India

Ribu Mathew

Manipal Academy of Higher Education (MAHE), Manipal, India

Seema Rajput and Ankur Beohar

VIT Bhopal University, Bhopal, India

5.1 INTRODUCTION

Silicon is one of the most used materials for MOS devices, used to design any digital and analog/RF circuits. Downscaling a device's dimension while at the same time enhancing its performance results in severe short channel effects (SCEs). To mitigate the SCE concerns, several new device architectures have been used, such FinFET, gate-all-around (GAA) FET, Junctionless FET Tunnel FET, NC FET, Nanowires, FeFET, Nanosheet, and Graphene FET shown in Figure 5.1(a) for the development of various electronics gadgets, and several channel materials have been realized and researched by the

                 DOI: 10.1201/9781032670270-5

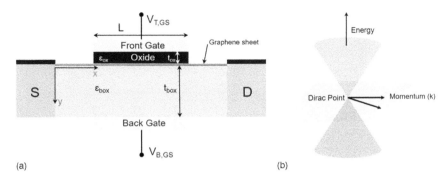

Figure 5.1 (a) Cross-sectional view of the Graphene FET, with drain (D), source (S), and back-gate (BG) contacts. (b) Energy and wave vector energy relation diagram of the monolayer graphene sheet.

scientific community [1–7]. These materials have shown their capabilities to maintain a proper trade-off between the device performance and SECs. Among these discovered materials, graphene has attracted enormous attention from the academic and scientific communities because of its extraordinary electrical and mechanical properties [8–14]. Because of the ultra-high mobility of charge carriers, a graphene-based FET device shown in Figure 5.1 produces very-high-speed electronics for a vast variety of RF applications.

However, its zero-bandgap shown in Figure 5.1(b) nature becomes the main hurdle for GFET for digital applications. It has ultra-high intrinsic charge carrier mobility and saturation velocity, which makes graphene-based MOS devices (GFET) compatible with high-speed analog/RF circuit applications. This chapter presents a detailed discussion of the different physical parameters that are used in the analytical modeling of graphene-based FET devices.

5.2 ELECTRICAL AND MECHANICAL PROPERTIES

Graphene is a monolayer atomic thick sheet of sp^2 hybridized carbon atoms having unique material properties regarding many electrical applications. Importantly, it has superior electrical-thermal conductivity, optical transparency, and chemical stability over other semiconducting materials. Also, it exhibits many other desirable properties including high thermal conductivity (5000 $Wm^{-1}K^{-1}$) and high critical current densities ($\approx 3 \times 10^9$ A/cm^2) [8, 9]. Further, the atomic thickness of graphene monolayer sheets is ideally utilized in the large area nanomanufacturing of wearable and flexible electronic devices.

In addition to electrical superiority, it also possesses excellent mechanical strength in terms of the strain limit of the material. Note that the strain limit

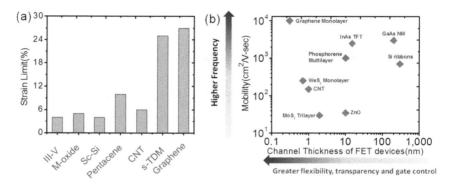

Figure 5.2 Comparison of the thicknesses vs. the carrier mobility of different atomic thin materials. The atomic thickness of 2D materials will offer better device electrostatic and gate, while the ultra-high mobility of atomic thin material may allow higher-frequency FET devices.

decreases linearly with the thickness of the materials as shown in Figure 5.2(a) and (b). Therefore, graphene has better mechanical strength than any other available semiconductor.

Apart from this, graphene also exhibits the highest carrier mobility of any known material, which is 100 times greater than Si and better than other high-mobility 2D semiconductors. Owing to the smaller effective mass of the charge carriers, graphene affords the highest charge carriers' mobility at relatively high carrier density (e.g., $10^{12}/cm^2$).

However, this carrier mobility in any MOS device only describes the charge carrier speed under a low electric field. While highly scaled FET devices have a very high field in the channel region, in this case the carrier saturation velocity (V_{sat}) becomes an important factor for the transport of charge carriers. The graphene exhibits ultra-high carrier saturation velocity, which is expected to reach around.

$4.5 \times 10^7 cm^{-1}$. In addition, under high fields, the V_{sat} in graphene does not drop as severely as those in III-V semiconductors. Therefore, the extraordinarily high carrier transport makes it a suitable material for ultra-high-speed flexible RF electronics [8–14].

5.2.1 E-k relationship

The E-k relationship describes the association between the energy and momentum of available quantum mechanical states for electrons in the semiconductor material.

The E-k relationship for graphene in the first Brillouin zone (BZ) is given by [12, 13]

$$E\left(k\right) = E - E_{CV} = s\hbar v_F \left|k\right| \tag{5.1}$$

where $s = +1$ for conduction band, $s = -1$ for valence band, and v_F and $\hbar$ are the Fermi velocity and reduced Plank's constant, respectively. $|k|(= {}^p k_x{}^2 + k_y{}^2)$ is the wave vector of the charge carrier in a two-dimensional (2D) mono-layer graphene sheet; $|k| = 0$. The point where the minima of CB are met with the maxima of VB is called Dirac Point. In graphene, the conduction band minima (E_C) and the valence band maxima (E_V) coincide with each other, i.e., $E_C = E_V = E_{CV}$. The E-k relation is directly proportional to the number of states per interval of energy at each energy level available to be occupied by the charge carrier.

5.2.2 Density of state

The density of state (DoS) is the number of states per interval of energy at each energy level available to be occupied. Hence, N is the number of available states in k space in the case of graphene material, considering the valley and spin degeneracy factor, is given by [12–15]

$$N = g_s g_d \frac{\pi k^2}{\left(2\pi/L\right)^2} = g_s g_d \frac{Ak^2}{4\pi} \tag{5.2}$$

or

$$N = g_s g_d \frac{A}{4\pi} \left(\frac{E - E_{CV}}{\hbar v_f} \right)^2 \tag{5.3}$$

where g_s is spin degeneracy, g_d is degeneracy valley factors, and A is the area of the 2D graphene sheet.

The DoS per unit area $D(E)$ for graphene is given as

$$D(E) = \frac{1}{A} \frac{dN}{dE} \tag{5.4}$$

Putting the of expression of N from Equation (5.3) into Equation (5.4), we have the DOS for graphene material:

$$D(E) = \frac{2}{\pi} \frac{|E - E_{CV}|}{\left(\hbar v_f\right)^2} \tag{5.5}$$

The density of the state is used to derive the expressions for the voltage-dependent hole and electron sheet carrier densities in the channel region of graphene FET.

5.2.3 Carrier concentration

The voltage-dependent carrier density of the hole in the graphene channel is given by [21]

$$p = \int_{-\infty}^{E_{CV}} D(E)\big[1 - f(E)\big]dE \tag{5.6}$$

where $f(E)$ is the Fermi Dirac distribution.

Now, putting DOS from Equation (5.5) into Equation (5.6), the hole carrier sheet density is given by

$$p = \int_{-\infty}^{E_{CV}} \frac{2}{\pi} \frac{|E - E_{CV}|}{(\hbar v f)^2}\big[1 - f(E)\big]dE \tag{5.7}$$

Now, changing the order of integration of the above equation can be written as

$$p = \int_{-\infty}^{E_{CV}} \frac{2}{\pi} \frac{|E + E_{CV}|}{(\hbar v f)^2}\big[1 - f(E)\big]dE \tag{5.8}$$

Using E_{CV} as a reference energy, i.e., for $E_{CV} = 0$, $E_f = q \times V_{ch}$ is the Fermi energy, and V_{Ch} is the channel potential drop across the graphene channel.

Taking into account the above equation, the hole density in the graphene sheet is given by

$$p = \frac{2}{\pi (\hbar v f)^2} \int_{0}^{\infty} \frac{E}{\exp\left(\dfrac{E + E_{CV}}{K_B T}\right) + 1} dE \tag{5.9}$$

In a similar way, the expression for electron sheet density in graphene can be given as

$$n = \frac{2}{\pi (\hbar v f)^2} \int_{0}^{\infty} \frac{E}{\exp\left(\dfrac{E - E_{CV}}{K_B T}\right) + 1} dE \tag{5.10}$$

where K_B is the Boltzmann constant and T is the temperature.

5.2.4 Channel charge

Now the net carrier density (Q_{net}) in the graphene sheet participating in the conduction of GFET is equal to the difference between hole sheet density and electron sheet density and is given as [21]

$$Q_{net} = q \times (p - n) \tag{5.11}$$

$$Q_{net} = \frac{2q}{\pi (\hbar v_f)^2} \left[\int_0^\infty \frac{E}{\exp\left(\dfrac{E + E_{CV}}{K_B T}\right) + 1} dE - \int_0^\infty \frac{E}{\exp\left(\dfrac{E - E_{CV}}{K_B T}\right) + 1} dE \right] \tag{5.12}$$

The above equation gives the net charge carrier density that participated in the current conduction of GFET. However, the carrier density near the Dirac point generated due to thermal excitation is still underestimated.

In zero-bandgap graphene material holes, electrons, and the residual charge carrier density concurrently participated in the overall current conduction in GFET.

The new equation for the sheet density in graphene FET is given as [15]

$$Q_{tot} = q \times (p + n) \tag{5.13}$$

or

$$Q_{tot} = \frac{q\pi (K_B T)^2}{3(\hbar v_f)^2} + \frac{q^3 V_{ch}^2}{\pi (\hbar v_f)^2} + q n_{pud} \tag{5.14}$$

where n_{pud} is electron-hole puddles generated by residual charge for a monolayer graphene sheet given by [16]

$$n_{pud} \approx \frac{2}{\pi \hbar^2 v_f^2} \left(\frac{\Delta^2}{2} + \frac{\pi^2}{6} K_B^2 T \right) \tag{5.15}$$

Upadhyay et al. [12] claimed the improved accuracy of the GFET models when residual charge density for small ($q V_{Ch} \ll K_B T$) and larger values ($q V_{Ch} \gg K_B T$) of V_{Ch} is taken into account. The channel carrier density is used to calculate the quantum and different parasitic capacitances formed at the different interfaces of Graphene FET by using the Mayer and the charge conservation approach. These approaches are described in detail in the next section.

5.2.4.1 Terminal charge

The charge distribution in the GFET with respect to the terminal voltages is required for the accurate modeling of the intrinsic capacitances in GFET devices. The charge at the top/bottom gate terminal has been obtained after applying Gauss' law to the top/bottom gate stack.

The resulting equations for Q_g, Q_b, Q_d, and Q_s are [17, 18]

$$Q_g = \frac{WC_t}{C_t + C_b}\left[C_b L\left(V_{gs} - V_{gs0} - V_{bs} + _{bs0}\right) - \int_0^L Q_{net}dx\right] \tag{5.16}$$

$$Q_b = \frac{WC_b}{C_t + C_b}\left[C_t L\left(V_{bs} - V_{bs0} - V_{gs} + _{gs0}\right) - \int_0^L Q_{net}dx\right] \tag{5.17}$$

$$Q_d = W\int_0^L \frac{x}{L}Q_{net}dx \Bigg] \tag{5.18}$$

$$Q_s = -\left(Q_g + Q_b + Q_s\right) \tag{5.19}$$

The above expressions can be conveniently used to formulate the different trans-capacitance associated with the terminals of the GFET device.

5.3 CAPACITANCE

In the case of Graphene FET, two types of approaches are used to calculate capacitances formed at the different interfaces of Graphene FET: (1) the Mayer technique and (2) the charge conservation techniques.

5.3.1 Mayer technique

Mayer's approach is the most-used approach because of its simplicity and fast computation. In this approach, it is assumed that the capacitances in the intrinsic FETs are reciprocal; the channel charge is given by [20, 21]

$$Q_{CH}\left(x\right) = W\int_0^L \left(Q_{net}\left(x\right) + qn_{pud}\right)dx \tag{5.20}$$

where symbols have their usual meanings.

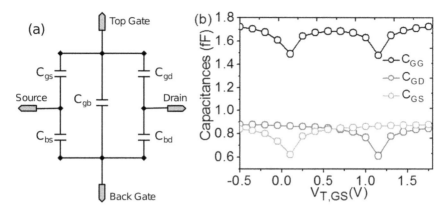

Figure 5.3 Comparison of the thicknesses vs. the carrier mobility of different atomic thin materials. The atomic thickness of 2D materials will offer better device electrostatic and gate, while the ultra-high mobility of atomic thin material may cause the higher frequency FET devices.

The above equation is solved by changing the dx into dV. By changing the limits of the integration we have a new equation given by

$$Q_{CH}(x) = \frac{qW}{E_{av}} \int_0^{V_{DS}} \left(Q_{net}(x) + qn_{pud} \right) dx \qquad (5.21)$$

where $E_{av} = \dfrac{dV}{dx} \approx \dfrac{V_{DS}}{L}$ and the remaining symbols have their usual meanings

5.3.1.1 Gate-source capacitance C_{gs}

The small-signal gate-source capacitance can be calculated as

$$C_{gs} = \frac{dQ_{CH}}{dV_{GS}} |_{V_{DS}=constt} \qquad (5.22)$$

5.3.1.2 Gate-drain capacitance C_{ds}

The small-signal gate-drain capacitance can be calculated as

$$C_{ds} = \frac{dQ_{CH}}{dV_{DS}} |_{V_{GS}=constt} \qquad (5.23)$$

The Mayer technique assumes that the capacitances in the intrinsic FETs are reciprocal in nature. It cannot fit in the real scenario because it cannot ensure charge conservation.

5.3.2 Ward-Dutton's techniques

Ward-Dutton's technique is based on ensuring charge conservation and incorporating the non-reciprocity capacitance property of MOSFET devices, which is exclusively required for RF circuits. The influence of transcapacitances is important and should be considered. The charge-based capacitance modeling for a 4-terminal MOS device can be done with 4 self-capacitances and 12 intrinsic transcapacitances, which makes 16 capacitances in total.

These 16 capacitances form the capacitance matrix, in which each element C_{ij} is described as the dependence of the charge at terminal i with respect to the varying bias voltage at terminal j, if the voltage is constant at another terminal [18–20]:

$$C_{ij} = -\frac{\partial Q_i}{\partial V_j} \text{ for } i \neq j \text{ and } C_{ij} = -\frac{\partial Q_i}{\partial V_j} \text{ for } i \neq j \tag{5.24}$$

where i and j stand for S, D, G, and B

$$\begin{bmatrix} C_{GG} & -C_{GD} & -C_{GS} & -C_{GB} \\ -C_{DG} & C_{DD} & -C_{DS} & -C_{DB} \\ -C_{SG} & -C_{SD} & C_{SS} & -C_{SB} \\ -C_{BG} & -C_{BD} & -C_{BS} & C_{BB} \end{bmatrix}$$

The elements of each row and column must sum to zero to preserve the charge conservation principle of the given device. Note that in 16 intrinsic capacitances, only 9 are independent; the rest are the dependent capacitances.

$$C_{DD} = \frac{\partial Q_D}{\partial V_D} = \frac{\partial Q_D}{\partial V_{CD}} \times \frac{\partial V_{CD}}{\partial V_D} \tag{5.25}$$

$$C_{DB} = \frac{\partial Q_D}{\partial V_B} = -\frac{\partial Q_D}{\partial V_{CD}} \times \frac{\partial V_{CD}}{\partial V_B} = -\frac{\partial Q_D}{\partial V_{CS}} \times \frac{\partial V_{CS}}{\partial V_B} \tag{5.26}$$

The following derivation, used to derive all capacitances, is given by

$$\frac{\partial V_{CS}}{\partial V_G} = \frac{\partial V_{CD}}{\partial V_G} = \frac{C_{Tox}}{C_{Box}} \frac{\partial V_{CS}}{\partial V_G} = \frac{C_{Tox}}{C_{Box}} \frac{\partial V_{CD}}{\partial V_G} = \frac{C_{Tox}}{C_{Tox} + C_{Box} + C_q \left(V_{Ch}(x) \right)} \tag{5.27}$$

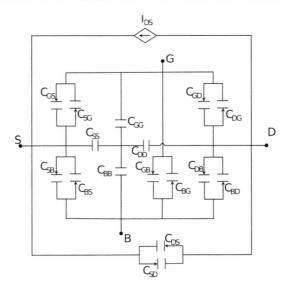

Figure 5.4 Ward-Dutton's equivalent intrinsic capacitive model of GFET device.

$$\frac{\partial V_{CS}}{\partial V_S} = \frac{\partial V_{CD}}{\partial V_D} = \frac{C_{Tox} + C_{Box}}{C_{Tox} + C_{Box} + C_q\left(V_{Ch}(x)\right)} \qquad (5.28)$$

$$\frac{\partial V_{CS}}{\partial V_S} = \frac{\partial V_{CD}}{\partial V_D} = 0 \qquad (5.29)$$

$\frac{\partial V_{CS}}{\partial V_D} = \frac{\partial V_{CD}}{\partial V_S} = 0$, because the voltage applied at the source terminal has no control over the charge at the drain terminal, and vice versa.

The following relations between the top and back-gate capacitances can be given as

$$C_{BD} = C_{GD}\left(\frac{C_{Box}}{C_{Tox}}\right) \qquad C_{DB} = C_{DG}\left(\frac{C_{Box}}{C_{Tox}}\right) \quad C_{BS} = C_{GS}\left(\frac{C_{Box}}{C_{Tox}}\right)$$

$$C_{SB} = C_{SG}\left(\frac{C_{Box}}{C_{Tox}}\right) \qquad C_{BB} = C_{GG}\left(\frac{C_{Box}}{C_{Tox}}\right)$$

$$C_{BG} = C_{GB} = C_{GG}\left(\frac{C_{Box}}{C_{Tox}}\right) = -C_{BB}\left(\frac{C_{Box}}{C_{Tox}}\right)$$

5.3.3 Quantum capacitance

The quantum capacitance in the GFET device is formulated by differentiating the net channel sheet charge density (Q_{sh}) with respect to channel voltage (V_{Ch}) [20],

$$C_q = -\frac{dQ_{net}}{dV_{Ch}} \tag{5.30}$$

The '-'ve sign in Equation (5.30) is due to the more positive at gate terminal will result in a more negative charge carriers in the channel region of GFET. This leads to a more negative charge in the channel region of the GFET device. After solving the differential equation and performing the mathematical manipulations, we have [21]

$$C_q = \frac{2q^2 K_B T}{+\left(\hbar v_f\right)^2} \ln\left[2\left(1 + \cos h\left(\frac{qV_{Ch}}{K_B T}\right)\right)\right] \tag{5.31}$$

where q is the elementary charge, K_B is the Boltzmann constant, T is the temperature, $\hbar$ is the reduced Plank's constant, v_f is the Fermi velocity, and V_{Ch} is the voltage drop across the channel. With condition $qV_{Ch} \gg K_B T$, the (5.31) is simplified in the form

$$C_q = \frac{2q^2}{\pi} \frac{q|V_{Ch}|}{\left(\hbar v_f\right)^2} \tag{5.32}$$

where symbols have their usual meanings.

5.4 CHANNEL POTENTIALS

The compact modeling for the drain current of Graphene FET has three steps: In the first step, the capacitance associated with the metal–insulator–graphene (MIG) structure is modeled; then the charge associated MIG structure is formulated; and finally the drain current equation [20].

5.4.1 Channel potential modeling

The formulation of charge in a MIG structure needs a different approach in comparison to conventional semiconductors-based FET devices, due to the zero bandgap of graphene material. For this purpose, we incorporate the specific DoSs of graphene with the Fermi approach for the carrier distribution.

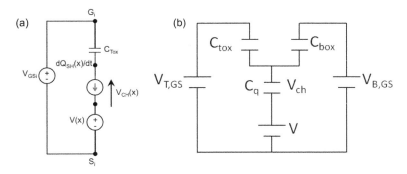

Figure 5.5 Equivalent circuit of the MIG capacitor.

After simplification of the mathematical expression, the total charge density in the MIG capacitor can be approximated as follows [20–22]:

$$Q_{SH}(x) \approx q\left(-\frac{q^2}{\pi(\hbar v_f)^2}|V_{CH}(x)|V_{CH}(x) + N_A - N_D\right) \tag{5.33}$$

$$= q\left(-\frac{\beta}{q}|V_{CH}(x)|V_{CH}(x) + N_f\right) \tag{5.34}$$

where q the electronic charge, $V_{CH}(x)$ the channel potential along channel length x, N_A is the acceptor, and N_D is the donor doping concentration, which is equal to the intrinsic carrier concentration, i.e., $N_A = N_D = N_f$ and $\beta = q^3/\pi(\hbar v_f)^2$.

Applying Kirchhoff's law to the equivalent circuit shown in Figure 5.5(b), the relation among the gate voltage, channel potential, and potential variation due to V_{DS} can be established [23]:

$$(C_{Tox})\left(V_{CH}(x) - V_{Gsi} + V(x) - \beta|V_{CH}(x)|V_{CH}(x) + qN_f\right) \tag{5.35}$$

The zeros of this second-degree polynomial can be calculated and give the following channel potential solution [24]:

$$V_{ch}(x) - \left(V_{T,eff} - \frac{V_{DS}}{2}\right) \times \frac{C_{tox}}{(C_{tox} + C_{box} + 0.5C_q)}$$
$$+ \left(V_{B,eff} - \frac{V_{DS}}{2}\right)\frac{C_{box}}{(C_{tox} + C_{box} + 0.5C_q)} \tag{5.36}$$

where C_{tox}, is the top gate oxide capacitance per unit area, while C_{box} is the back gate oxide capacitance per unit area. $C_{tox} = \epsilon_{tox, box}/t_{ox}$, where $t_{tox,box}$ is the thickness of the top, back-gate oxide layer and $\epsilon_{tox, box}$ is the permittivity of the top and back oxide material. $VT,eff = (VTGS - VTGS,0)$ and $VB,eff = (VBGS - VBGS,0)$ are the top and back-gate overdrive voltages, respectively. C_q is the quantum capacitance.

5.5 DRAIN CURRENT MODELING OF GRAPHENE FET: DRIFT-DIFFUSION THEORY

In this section, the modeling of drain current (I_{DS}) is based on the drift-diffusion approach for single-layer GFET. Figure 5.1(a) shows the cross-sectional view of the model transistor. It consists of a large-area graphene monolayer as a channel material located on a thick dielectric substrate that is grown on a heavily doped Si wafer.

In general, the drain current I_{DS} of a field-effect transistor can be expressed as [25]

$$I_{DS} = -q\rho_{sh}(x)v(x)W \tag{5.37}$$

where $\rho_{sh}(x)$ and $v(x)$ are, respectively, position (x) dependent free carrier sheet density and carrier drift velocity in the channel region. The Monte Carlo simulations have shown the soft saturation in the drift velocity of graphene, which can be approximated by the expression

$$v(x) = \frac{\mu E}{1 + \dfrac{\mu E}{v_{sat}}} \tag{5.38}$$

where $E = dV(x)/dx$ is the electric field, μ is the carrier low-field mobility, and v_{sat} is the saturation velocity. The saturation velocity depends on the channel carrier concentration as

$$v_{sat} = \frac{\omega}{(\pi\rho_{sh})^{0.5+AV^2(x)}} \tag{5.39}$$

Now, combining Equations 5.37, 5.38, and 5.39, the drain current becomes

$$I_{DS} = -q\rho_{sh} \frac{\mu(-dV/dx)}{1 + \dfrac{\mu(-dV/dx)}{v_{sat}}} \tag{5.40}$$

Equation (5.40) can be solved by using the separation of variables integration method. The left-hand side of Equation (5.40) is integrated with respect to the x, with limits from $x = 0$ to $x = L$, while the right-hand side is integrated over the limit from $V(0) = 0$ to $V(L) = V_{DS}$. The final expression for the drain current equation is given by [20, 25]

$$I_{DS} = q\mu W \frac{\int_0^{V_{DS}} \rho_{sh} dV}{L + \mu \left| \int_0^{V_{DS}} \left(\frac{1}{V_{sat}} \right) dV \right|} \tag{5.41}$$

where $\rho_{sh} = (|Q_{net}| + q n_{puddle})$ is the free carrier sheet density in the channel and $Q_{net} = (Q_{SH} - qN_f)$ is the net bias–dependent mobile charge carrier. The detailed drain current equation of GFET is given as [22]

$$I_{DS} = \frac{\mu W \int_0^{V_{DS}} \left(|Q_{net}| + q n_{puddle} \right) dV}{L + \mu \left| \int_0^{V_{DS}} \frac{1}{v_{sat}} dV \right|} \tag{5.42}$$

where μ is the mobility, L and W are, respectively, the length and width of the graphene monolayer, $Q_{net} = q \times (p - n)$ is the net mobile charge density per unit area, p is the acceptor carrier density, n is the donor carrier density, q is the elementary charge ($1.6 \times 10^{-19}C$), v_{sat} is the saturation velocity, $n_{puddle} = \Delta^2/(\pi \hbar^2 v_f^2)$, and Δ represents the spatial in-homogeneity of the electrostatic potential.

For simplicity, the I_{DS} can be given by [21, 25]

$$I_{DS} = \mu W \frac{Num_1 + q n_{pub}}{Den} \tag{5.43}$$

Accordingly, Equation (5.43) can be rewritten as

$$Num_1 = \beta \int_0^{V_{DS}} \left[\frac{-C_{tox}}{2\beta} + \frac{\sqrt{C_{tox}^2 + 4\beta |C_{tox} (V_{eff} - V)|}}{2\beta} \right]^2 dV \tag{5.44}$$

Equation (5.44) can be simplified by using the integration variable to $z = C_{tox}(V_{eff} - V)$, and has the following solutions:

$$Num_{1(z>0)} = -\frac{1}{\beta^2 C_{tox}} \times \left[\frac{C_{tox}^4}{32} - \frac{C_{tox} \left(C_{tox}^2 + 4\beta z \right)^{3/2}}{12} + \frac{\beta^2 z^2}{2} + \frac{\beta C_{tox}^2 z}{2} \right]_{z_1}^{z_2} \tag{5.45a}$$

$$Num_{1(z<0)} = -\frac{1}{\beta^2 C_{tox}} \times \left[-\frac{C_{tox}^4}{32} - \frac{C_{tox}\left(C_{tox}^2 + 4\beta z\right)^{3/2}}{12} - \frac{\beta^2 z^2}{2} + \frac{\beta C_{tox}^2 z}{2} \right]\Bigg|_{z_1}^{z_2}$$

(5.45b)

where $z_1 = C_{tox} V_{eff}$ and $z_2 = C_{tox}(V_{eff} - V_{DSi})$.

The denominator can be expressed as

$$Den = L + \mu \left| \int_0^{V_{DS}} \frac{1}{v_{sat}} dV \right|$$

(5.46)

which can be simplified assuming an average v_{sat} given by

$$v_{sat,av} = \frac{\omega}{\sqrt{\pi \dfrac{|Q_{net,av}|}{q} + n_{pud}}}$$

(5.47)

where ω is calculated from the surface phonon energy of the substrate ($= \bar{}h\omega$) and $Q_{net,av}$ is the net average charge, which participated in the overall current conduction is given by

$$Q_{net,av} = \beta \left[\frac{-C_{tox}}{2\beta} + \frac{\sqrt{C_{tox}^2 + 4\beta \left| C_{tox}\left(V_{eff} - V_{DS}/2\right)\right|}}{2\beta} \right]^2$$

(5.48)

The simplified form of the denominator in Equation (5.42) can be expressed as

$$Den = L + \frac{\mu}{v_{sat,av}} |V_{DS}|$$

(5.49)

Replacing the technology-dependent parameters taken from the measured GFETs and physical constants unveils that there is a term that dominates, and therefore Equation (5.44) can be reduced to

$$Num_1 \simeq -\frac{1}{2} \frac{z^2}{C_{tox} \times sign(z)}\Bigg|_{z_1}^{z_2}$$

(5.50)

for $z > 0$ (5.50) can be expressed as

$$Num \simeq C_{tox} V_{DS} \left(V_{eff} - \frac{V_{DS}}{2} \right) \tag{5.51}$$

Now, Equation (5.49) can be simplified under the assumption that $V_{eff} > V_{DS}/2$ and $n_{pub} << \pi |Q_{net,av}|/q$,

$$Deb \simeq L + \frac{\mu}{\omega} \sqrt{\frac{\pi C_{tox}}{q}} V_{DS} \sqrt{V_{eff} - \frac{V_{DS}}{2}} \tag{5.52}$$

Finally, the GFET drain current is formulated by putting Equations (5.51) and (5.52) into Equation (5.43):

$$I_{DS} \simeq \frac{\mu W C_{tox} \left(V_{eff} - V_{DS}/2 \right)}{\dfrac{L}{V_{DS}} + \dfrac{\mu}{\omega} \sqrt{\dfrac{\pi C_{tox}}{q}} \sqrt{V_{eff} - V_{DS}/2}} \tag{5.53}$$

which is a closed analytical expression that relates the main technology parameters and biasing conditions.

5.6 SUMMARY

This chapter has presented a brief introduction to the electronic properties of graphene materials followed by the fundamental parameters of graphene. The E-k relation, density of state, channel concentration, terminal capacitances, quantum capacitance, and channel voltage are the parameters that are used in the analytical model development for GFET devices presented. Further, a brief discussion of different drain current modeling approaches has also been presented.

ACKNOWLEDGMENT

The authors would like to acknowledge the SERB TARE GRANT Project no. TAR/2022/000406, Govt. of India, for technical and financial support.

REFERENCES

[1] W. Arden, M. Brillout, P. Cogez, M. Graef, B. Huizing, and R. Mahnkopf, "More than Moore -White Paper," [Online]. Available: http://www.itrs2.net/

[2] A. Dutt, S. Tiwari, A. K. Upadhyay, R. Mathew, and A. Beohar, "Impact of drain underlap and high bandgap strip on cylindrical gate all around tunnel FET and its influence on analog/RF performance," *Silicon*, 2022.

[3] R. Mathew, A. Beohar, and A. K. Upadhyay, "High-performance tunnel field-effect transistors (TFETs) for future low power applications," *Semiconductor Devices and Technologies for Future Ultra Low Power Electronics*, pp. 29–57, 2021.

[4] Y. S. Song, S. Tayal, S. B. Rahi, J. H. Kim, A. K. Upadhyay, and B.-G. Park, "Thermal-Aware IC Chip Design by Combining High Thermal Conductivity Materials and GAA MOSFET," *2022 5th International Conference on Circuits, Systems and Simulation (ICCSS)*, May 2022.

[5] S. Tayal et al., "Incorporating bottom-up approach into device/circuit co-design for SRAM-based cache memory applications," *IEEE Transactions on Electron Devices*, pp. 1–6, 2022, doi:10.1109/ted.2022.3210070

[6] S. B. Rahi, S. Tayal, and A. Kumar Upadhyay, "Emerging negative capacitance field effect transistor in low power electronics," *Microelectronics Journal*, 116, 2021, 105242, https://doi.org/10.1016/j.mejo.2021.105242

[7] A. K. Upadhyay, S. B. Rahi, S. Tayal, and Y. S. Song, "Recent progress on negative capacitance tunnel FET for low-power applications: Device perspective," *Microelectronics Journal*, p. 105583, 2022.

[8] Y. Obeng, and P. Srinivasan, "Graphene: Is it the future for semiconductors? An overview of the material, devices, and applications," *Interface Magazine*, vol. 20, pp. 47–52, 2011.

[9] F. Schwierz, "Graphene transistors: Status, prospects, and problems," *Proceedings of the IEEE*, vol. 101, pp. 1567–1584, 2013.

[10] K. S. Novoselov, V. I. Falko, L. Colombo, P. R. Gellert, M. G. Schwab, and K. Kim, "A roadmap for graphene," *Nature*, vol. 490, pp. 192–200, 2012.

[11] Y. Wu, D. B. Farmer, F. Xia, and P. Avouris, "Graphene electronics: Materials, devices, and circuits," *Proceedings of the IEEE*, vol. 101, pp. 1620–1637, 2013.

[12] A. K. Upadhyay, A. K. Kushwaha, D. Gupta, and S. K. Vishvakarma, "Recent development in analytical model for graphene field effect transistors for RF circuit applications," arXiv:2101.01955.

[13] Abhishek Kumar Upadhyay, Ajay K. Kushwaha, S. K. Vishvakarma, "Compact modeling of graphene field effect transistors for RF circuit applications," PhD Thesis, 2019.

[14] S. A. Thiele, J. A. Schaefer, and Frank Schwierz. "Modeling of graphene metal-oxide-semiconductor field-effect transistors with gapless large-area graphene channels," *Journal of Applied Physics*, vol. 107, no. 9, p. 094505, 2010.

[15] Gerhard Martin Landauer, David Jimenez, and Jose Luis Gonzalez. "An accurate and Verilog-A compatible compact model for graphene field-effect transistors," *IEEE Transactions on Nanotechnology*, vol. 13, no. 5, pp. 895–904, 2014.

[16] Zhenxing Wang, Zhiyong Zhang, Huilong Xu, Li Ding, Sheng Wang, and LianMao Peng. "A high-performance top-gate graphene field-effect transistor based frequency doubler," *Applied Physics Letters*, vol. 96, no. 17, p. 173104, 2010.

[17] David Jiménez, "Explicit drain current, charge and capacitance model of graphene field-effect transistors." *IEEE Transactions on Electron Devices*, vol. 58, no. 12, pp. 4377–4383, 2011.

[18] Francisco Pasadas, and David Jiménez. "Large-signal model of graphene field-effect transistors—Part I: Compact modeling of GFET intrinsic capacitances," *IEEE Transactions on Electron Devices*, vol. 63, no. 7, pp. 2936–2941, 2016.

[19] D. Ward and R. Dutton "A charge-oriented model for MOS transistor capacitances," *IEEE Journal of Solid State Circuits*, vol. SC-13, pp. 703–708, 1978.

[20] S. Thiele, J. Schaefer, and F. Schwierz, "Modeling of graphene metal-oxidesemiconductor field-effect transistors with gapless large-area graphene channels," *Journal of Applied Physics*, vol. 107, pp. 094505-1–094505-8, 2010.

[21] Abhishek Kumar Upadhyay, Nitesh Chauhan, and S. K. Vishvakarma, "A Compact Electrical Modelling for Top-Gated Doped Graphene Field-Effect Transistor," *IETE Journal of Research*, vol. 64, no. 3, pp. 317–323, 2018. DOI :10.1080/03772063.2017.1355752

[22] A. K. Upadhyay, A. K. Kushwaha, and S. K. Vishvakarma, "A Unified Scalable Quasi-Ballistic Transport Model of GFET for Circuit Simulations," *IEEE Transactions on Electron Devices*, vol. 65, no. 2, pp. 739–746, 2018. doi:10.1109/TED.2017.2782658

[23] A. K. Upadhyay, A. K. Kushwaha, P. Rastogi, Y. S. Chauhan, and S. K. Vishvakarma, "Explicit model of channel charge, backscattering, and mobility for graphene FET in quasi-ballistic regime," *IEEE Transactions on Electron Devices*, vol. 65, no. 12, pp. 5468–5474, 2018. doi:10.1109/TED.2018.2877631

[24] A. K. Upadhyay, D. Gupta, R. Mathew, and A. Beohar, "A compact model of the backscattering coefficient and mobility of a graphene FET for SiO_2 and h-BN substrates," *Journal of Computational Electronics*, 2022.

[25] S. Rodriguez et al., "A comprehensive graphene FET model for circuit design," *IEEE Transactions on Electron Devices*, vol. 61, no. 4, pp. 1199–1206, 2014.

Chapter 6

Design of CNTFET-based ternary logic flip-flop and counter circuits using unary operators

Trapti Sharma

VIT Bhopal University, Bhopal, India

6.1 INTRODUCTION

Dimensional scaling is a suitable technique that provides promising results in improving the device density with enhanced performance. But with the reduction in the lateral dimension of the devices and interconnects, an increase in the leakage current further avoids the threshold voltage reduction, and device reliability is badly affected due to internally generated high electric fields. Other problems caused due to scaling are mobility degradation, gate-induced drain leakage, and threshold voltage fluctuation due to random dopants [1]. As there is a massive increase in compact smart electronic devices, energy savings or low battery consumption are the major concerns in VLSI circuit designs. Also, more than 70% of the chip area constitutes an interconnect, resulting in a major source of power dissipation for any digital design. Therefore, it is required to find some suitable radix that can reduce this interconnect power and provide minimal cost in performing digital circuit design.

One of the possible ways is to utilize some emerging device technology such as CNT technology and adopt multi-valued logic (MVL) design for digital computation. Unlike binary logic, in MVL design, arithmetic and logical operations can be carried out on more than two authorized levels [2]. Among various MVL systems, logic implementation using radix 3 is more economical, as it is nearest to the optimal exponent radix. In a ternary logic design that uses three logic levels, an enhanced amount of information is stored at each logic level in contrast to binary logic. As a result, the interconnect area is lowered as the number of signals used to represent the given range of numbers also gets decreased. Also, it is possible to process more information over the same set of wires or for a given register length, which in turn results in the reduction of circuit complexity for any logic realization. In ternary logic realization, fewer bits needed to realize the increased number of processing states results in the benefits of fast computation, dense data storage, and optimized connection routes. Thus it helps us to achieve power optimization due to interconnect complexity in digital design [2, 3].

DOI: 10.1201/9781032670270-6

The desirable characteristic to construct voltage operated MVL circuits is to have the device capability to operate with multiple threshold voltages. Some of the design techniques used to realize multiple threshold voltages in CMOS designs are adiabatic switching, dynamic threshold voltage control through biasing the CMOS wells, adjusting W/L ratios for the transistors, or adopting multiple supply voltages. Several problems arise, however, due to insufficient characteristics of MOS devices in realizing the multiple threshold voltages while creating MVL designs. It implies that digital circuit design using MOS devices does not provide a flexible and easy method to realize additional voltage levels in MVL design. Most of the MOSFET-based ternary logic designs utilize large off-chip resistors, multiple power sources, and supply voltages higher than the device threshold for realization [4].

Among various emerging devices, CNTFETs could be a possible alternative to conventional MOS devices, as threshold voltage variation can be performed by introducing changes in CNT diameter values. In other words, the ternary logic design involves the requirement of multiple threshold voltages that can be achieved by employing CNTs with different diameters. Also, it has low power overhead due to 1D ballistic operation and high current-carrying capability [5, 6].

Power consumption is an important performance metric for portable electronic devices. The power contribution of a digital design can be of static or dynamic type. As a matter of fact, within a circuit design, a short circuit power is 5–10% of dynamic power. The major component of dynamic power is due to various voltage level transitions. Counters are primary blocks in several VLSI systems such as A-to-D converters, D-to-A converters, frequency dividers, registers, etc. As counters are clock operated, designing them with low power overhead and high reliability is a major concern. The major contribution of power in counter designs is due to the clock input, which is acting as a driving signal for the flip-flops during their operation. This component is present even when flip-flops are not in the active state, resulting in additional power overhead. Thus, it is observed that, on average, 30–40% of the total power of sequential block design is due to clock power [7]. Hence high information-processing capability of ternary logic can be utilized here for the counter design. For instance, mod-16 counter realization using binary logic requires four flip-flops with equivalent clock signals. On the other hand, the same counter design using ternary logic requires three flip-flops with three clock input signals. Hence, power contribution can be reduced by exploiting the high computation capability of ternary logic.

Several works have reported CNTFET-based ternary logic circuits. Arithmetic circuits using different logic families are presented in [8–14]. Three-valued sequential cells such as D-flip-flop and counter utilizing successor and predecessor operators are presented in [15]. Here D-flip-flops and counter designs are constructed using static diode topology-based predecessor and successor elements. The design [15] provides high-performance metrics but has high static power consumption for logic 1 transitions. Some other D-flip-flop and counter designs are reported in [16–18] using other logic families. Hence, this work explores the benefits of ternary logic design techniques in realizing flip-flop and counter cells exploited for power and energy improvements by reducing interconnect complexity.

The chapter is structured as follows: The details of ternary unary operators such as inverters and shifting operators and CNTFETs are discussed in Section 6.2. The design details of the proposed cells are covered in Section 6.3. A discussion related to simulation results for performance evaluation is mentioned in Section 6.4. A conclusion follows.

6.2 PRELIMINARY

CNTFET is another variant of CMOS transistor whose channel material is a single or array of SWCNTs, as opposed to bulk silicon as in standard MOSFET construction. The CNTFET threshold voltage is approximatively derived as the half bandgap using Equation (6.1),

$$V_{TH} = \frac{E_g}{2e} = a.\frac{V_{pi}}{\sqrt{3}.e.D_{CNT}} \approx \frac{0.43}{D_{CNT}(nm)}(V) \tag{6.1}$$

The diameter of CNT is designated as D_{CNT}, which has a dependency on the chirality vector (n_1, n_2) and is given by Equation (6.2):

$$D_{CNT} = \frac{\sqrt{3}.a_0.\sqrt{n_1^2 + n_2^2 + n_1.n_2}}{\pi}(nm) \tag{6.2}$$

According to Equation (6.2), the threshold voltage decreases with increasing CNT diameter, and the chirality vector or wrapping vector (n1, n2) increases with increasing CNT diameter [5, 8]. For example, CNTFETs with a chirality vector of $(n_1, n_2) = (19, 0)$ have a CNT diameter of 1.487 nm and a threshold voltage of 0.293 V. Apart from that, CNTFET's features of increased quantum capacitance, enhanced temperature susceptibility, and reduced leakage power in sub-10nm dimensions make them suitable for deeply scaled technologies as compared to conventional MOS devices, and they can be employed in the designing of advanced digital designs. The ability to build MVL designs, particularly ternary circuits, is made possible by the CNT device's novel property of achieving multiple-threshold voltages by varying CNT diameter values.

Owing to the similar characteristics of MOSFET-type CNTFET with MOSFET in terms of the principle of operation and other advantageous features, in this work M-CNTFET is employed for designing and simulation of the proposed sequential blocks.

Table 6.1 Truth table of unary operators

Input(A)	NTI(A)	STI(A)	PTI(A)	A_1	Single shift(A(+1))	Dual shift(A(+2))
0	2	2	2	0	1	2
1	0	1	2	2	2	0
2	0	0	0	0	0	1

Ternary logic uses three logic levels, i.e., logic 0, 1, and 2, to realize digital functions at the voltage levels of 0, $V_{dd}/2$, V_{dd}, respectively. Table 6.1 illustrates three forms of inversion in ternary logic: a negative ternary inverter, a positive ternary inverter, and a standard ternary inverter. Apart from inversion, other unary operators are single-shift and dual-shift operators [11], which are used for the D-flip-flop designs and operations as depicted in Table 6.1.

6.3 PROPOSED DESIGN

Counter and registers are the primary elements in sequential block design in processor applications. In this section suggested flip-flop and counter designs are constructed by using the rotational symmetry of the shifting operators. At first, by taking an average of unary literal shifting operators, single shift and dual shift are realized. After that D flip-flops are designed based on a master-slave scheme involving two latch circuits. Further, these flip-flop designs are combined to construct the counter designs. As compared to binary counters, ternary counters provide the benefits of reduced interconnect complexity, which results in reduced chip area. Also, designing synchronous counter cells in which each flip-flop is fed with clock input involves more frequent transitions. In such a case, ternary logic enhanced processing capability results in reduced power overhead due to clock transitions. In other words, for mod-16 synchronous counter design, four clock inputs need to be applied, while for the ternary case, only three clock input signals are needed. Hence this high computational capacity of ternary logic can be exploited for the realization of dense and power-efficient sequential designs.

6.3.1 Proposed unary operators

As the implementation of the proposed flip-flop and counter designs is carried out using shifting literals, novel single-shift and dual-shift operators design is explained first. The operations of single shift and dual shift can be derived from Table 6.1 as inputs being shifted to their next logic level and inputs being shifted to their prior logic level, respectively. Therefore, the same logic may be reached when these single-shift and dual-shift blocks are connected in a cascade.

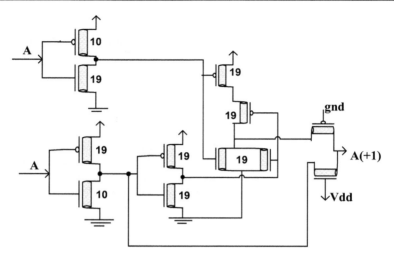

Figure 6.1 Proposed single-shift circuit using CNTFETs.

The circuit realization of CNTFET-based successor or single-shift operator is depicted in Figure 6.1. The number beside each transistor indicates the diameter values that are taken into account for circuit design. For single-shift operator realization, on average, two unary operators, as opposed to PTI and A_1 literal, are applied to a voltage divider circuit. The voltage divider circuit in use here offers low power usage for the specified designs. The basic reasoning adopted is that P-type and N-type transistors are not capable of passing strong values of ground and V_{dd} voltage levels. So ON-current of this always-on transistor will get decreased in the voltage division process, which leads to a reduction in the overall power dissipation of the circuit. The reduction of power components in voltage division results in overall energy conservation when creating single-shift operator designs. Since multi-stage counter design and D-flip-flop implementation use these unary operator blocks multiple times, overall power consumption of the derived counter cells also gets improved. Likewise, for the realization of the dual-shift operator design, an average of NTI inputs and $\overline{A_1}$ is applied to the voltage divider circuit and dual-shift output is obtained as shown in Figure 6.2.

The proposed designs of shifting operators are different as compared to existing ones [10, 11, 14]. Earlier works such as shifting circuit design [10] employ a smaller number of transistors, but an additional supply voltage of $\dfrac{V_{dd}}{2}$ is required for the implementation. Also, the circuit has less driving capability when used for designing complex circuits such as ripple carry adder designs. Other shifting operator designs are presented in [11]. The circuit is realized using a static diode configuration for voltage division. However, that design has high power consumption because a direct path exists between the voltage source and the ground, especially for realizing logic 1 transition. Similarly, other works have designed shifting circuits

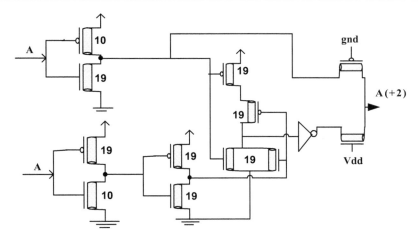

Figure 6.2 Proposed dual-shift circuit using CNTFETs.

using static diode configuration and utilizing different transistor arrangements. As compared to earlier works [10, 11, 15, 18], the proposed shifting designs are having improved performance due to efficient voltage divider topology.

6.3.2 Proposed D-flip-flop design

In conventional flip-flop designs, inverters are the primary elements for the realization. In this work, however, two shifting operators are used to generate the D-flip-flop outputs. As depicted in Figure 6.3(a), the operation of a D-flip-flop can be described as follows: on every clock transition, the input provided to the D-input is sent to the output (Q).

Initially, D-latch circuit is designed by cascading proposed single-shift and dual-shift operator circuits with appropriate clock signals as shown in Figure 6.3(b). Here when D-input is fed to a single-shift circuit after passing through a transmission gate controlled by the clock signal, the input is shifted to the next logic level. After that, this input is fed to a dual-shifting circuit, which in turn shifts the earlier output back to its previous logic level. For instance, suppose applied input D is at logic 0, then after passing through the single-shift circuit, the output appears is at logic 1. Further, when this logic 1 input is applied to the dual-shift circuit, the output of logic 0 appears across the latch output. Likewise, for other logic inputs, the latched outputs can be obtained. Here this rotational property is utilized for the implementation of D latches. Further D flip-flops are constructed by combining the D-latches in a master-slave configuration. So whenever the clock is at one logic (low logic), the input applied (D) is passed to the master stage output, and during that transition the slave stage is in OFF condition. Later if the clock is at another logic (high logic), the slave becomes active, passing latched output of the master stage further to the slave stage output.

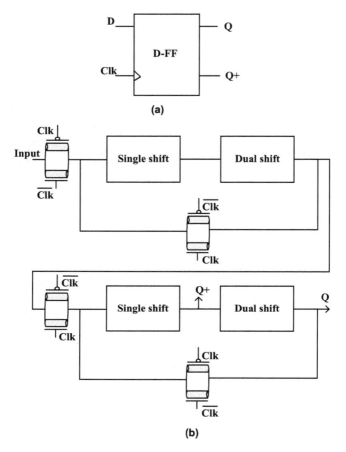

Figure 6.3 Proposed D-flip-flop design. (a) Block diagram; (b) Proposed D-flip-flop realized using CNTFET shifting operators.

6.3.3 Ternary counter design

Some of the common applications where these flip-flop designs are employed are counter and registers within a processor chip. Further, to check the applicability of proposed flip-flops in realizing compact structures, they are combined to realize asynchronous counter structures. In asynchronous counters, the clock input is applied to the first stage, and further stages inputs are derived from the earlier stage output. Asynchronous counters are having a smaller number of clock input signals as compared to synchronous ones. However, the propagation delay for the same grows to a greater extent when the input from the first stage is provided as the clock input for the next stage.

Figure 6.4(a) depicts a block diagram of the suggested three-stage asynchronous up counter, and Figure 6.4(b) depicts a 3-bit down counter. The basic

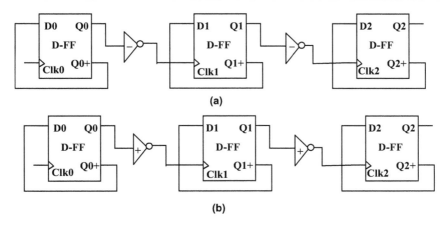

Figure 6.4 Proposed ternary 3-bit up-counter and down-counter cells realized using D flip-flops.

reasoning considered for the counter implementation can be obtained from the operation of mod-8 counter design as indicated in Table 6.2. As it is clearly observed in the first stage, to obtain the next-stage output, the feedback input that should be fed is Q(+1) input or a single shifted version. Further, second- and third-stage clock inputs are produced by taking NTI of whatever the outputs received from the previous stage for up-counter structures.

It is worth noting here that the proposed counter designs are more power efficient and have better energy efficacy as compared to the latest counter designs [15]. This is because the voltage divider topology adopted to realize the proposed design is having less current driving power as compared to existing ones. Thus, in the proposed work, lower power consumption and reduced interconnect complexity of ternary logic are leveraged in implementing sequential designs in processor systems.

Table 6.2 Truth table of three-stage ternary counter

Present state		Next state		Inputs	
Q1	Q0	Q1	Q0	D1	D0
0	0	0	1	0	1
0	1	0	2	0	2
0	2	1	0	1	0
1	0	1	1	1	1
1	1	1	2	1	2
1	2	2	0	2	0
2	0	2	1	2	1
2	1	2	2	2	2
2	2	0	0	0	0

6.4 RESULTS AND DISCUSSION

The performance assessment and comparison of the suggested design with other current works are described in this section. All simulations are performed using the Synopsis HSPICE simulator and the Stanford 32 nm CNTFET model. This standard model [19, 20] is designed for enhancement-mode unipolar MOSFETs like CNTFETs, in which each transistor's channel consists of one or more CNTs. To make the device compatible with the CMOS process, CNTFET circuits in the specified model [20] utilize the normal metal connection technique that has been used for silicon technology. Table 6.3 outlines several of the most significant simulation parameters.

Further, for the input setup, input supply voltage V_{dd} of 0.9 V, 1fF load capacitor, and room temperature conditions are considered. The analysis of the circuit is done in terms of performance parameters, including average power consumption, propagation delay, power delay product (PDP), and the number of transistors. For the calculation of power consumption, input patterns covering various logic levels are applied across the designs. For the propagation delay, the calculation is done for all possible transitions such as 0 to 1, 1 to 2, logic 2 to logic 0, and so on. After that, the maximum of all those is taken as the worst-case delay. Further, to find a trade-off, the power delay product is evaluated to make a fair comparison.

The suggested flip-flop and counter designs are implemented using shifting operators, hence an initial comparison of unary operators is done. Table 6.4 shows the comparison of shifting operator's as opposed to single shift and dual shift in terms of circuit parameters. The tabular findings demonstrate that power and energy savings of 44.4% and 60.19%, respectively, are obtained with the suggested single-shift circuit, while improvements of 69.20% and 74%, respectively, are obtained with the dual-shift design, compared to the design in [11, 15]. This improvement results from using a voltage divider topology that is more efficient than previous ones to build unary operators. As these are the basic circuits utilized further to construct D flip-flops and counter designs, it affects the performance of overall module.

Table 6.3 CNTFET parameters taken for simulation

Parameters	Description	Value
L_{ch}	Physical channel length	32 nm
L_{geff}	Mean free path in the intrinsic CNT channel	100 nm
L_{ss}	Length of doped CNT source-side extension region	32 nm
L_{dd}	Length of doped CNT drain-side extension region	32 nm
K_{gate}	Dielectric constant of high-k top gate dielectric material	16
T_{ox}	Thickness of high-k top gate dielectric material	4 nm
C_{su}	Coupling capacitance between the channel region and substrate	40 pf/m
E_{fi}	The Fermi level of the doped S/D tube	0.6 eV

Table 6.4 Simulation results of unary operator designs

	Power (uW)	Delay (ps)	PDP (aJ)	# transistors
Single shift				
Design [11]	0.6623	41.33	27.372	9
Design [15]	0.3172	40.18	12.745	8
Proposed	0.1767	28.199	4.982	12
Dual shift				
Design [11]	0.6273	35.12	22.030	9
Design [15]	0.5138	50.17	25.773	8
Proposed	0.1932	28.952	5.593	14

Further ternary D flip-flops are implemented using the rotational property of shifting operators applied with suitable clock signals. Table 6.5 depicts the comparison of various CNTFET-based ternary D-flip-flop designs. The results suggest a 22% and 53% improvement in power usage and PDP, respectively, compared to the least reported PDP of the design in [15]. The transient response is also displayed in Figure 6.5 to confirm the proposed flip-flop design's functionality. The inputs applied are clock (clk) and D input signals while the output obtained is Q. As it is shown clearly on every rising edge of the clock input signal, the input D is passed to the output Q.

Furthermore, three-stage up-counter and down-counter structures are realized using the proposed D-flip-flop cells. Table 6.6 compares the performance of multi-stage counter designs realized a maximum up to three stages in terms of primary design metrics. The outcomes reveal a PDP and power consumption decrease of 17% and 35%, respectively, for the 3-bit up-counter design and 15% and 33%, respectively, for the 3-bit down-counter design. Figure 6.6 illustrates the input/output waveforms of the three-stage counter to validate the functionality of suggested counter designs. In the figure, Q0, Q1, and Q2 indicate the corresponding outputs of stages 1, 2, and 3.

Hence it is worth noting that by using improved voltage divider topology in realizing proposed designs, power and energy enhancements are achieved as compared to recent works. Therefore, multi-valued logic design can be utilized to investigate the advantages of having greater computational capacity and to achieve power and energy improvements in the sequential blocks for processor design.

Table 6.5 Simulation results of ternary D-flip-flop circuits

D flip-flop	Power (uW)	Delay (ps)	PDP (aJ)	Set time (ps)	Hold time (ps)
Design[16]	1.290	35.00	45.15	10.0	5.0
Design[15]	0.5390	38.477	20.739	5.560	2.030
Proposed	0.42193	23.164	9.772	5.718	2.785

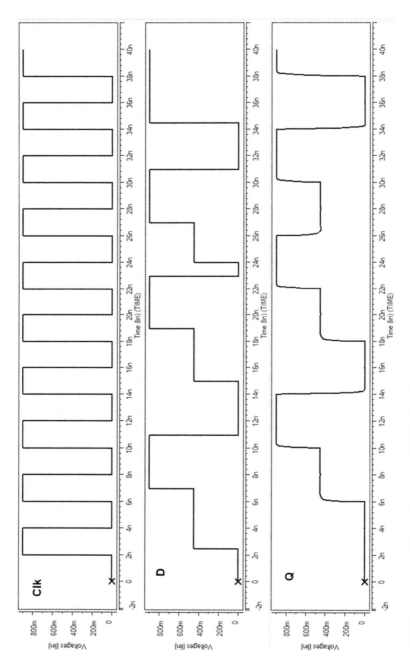

Figure 6.5 Transient waveform of proposed D-flip-flop design.

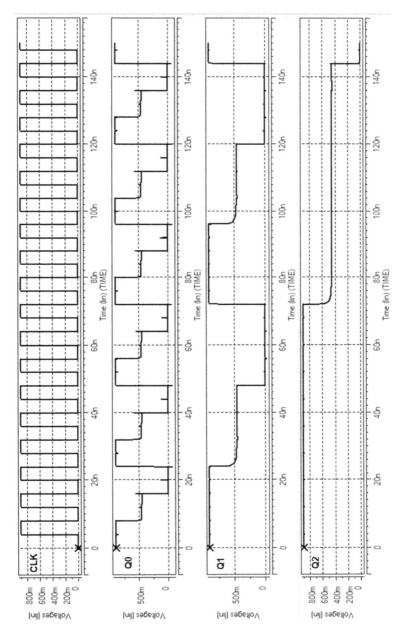

Figure 6.6 Transient response of 3-bit proposed ternary counter design.

Table 6.6 Simulation results of ternary up- and down-counter designs

Counter	No. of stages	Power (uW)	Delay (ps)	PDP (aJ)
Design [15]	1	0.4010	51.762	20.756
	2	0.7555	71.652	54.133
	3	1.0598	140.31	148.700
Proposed (up)	1	0.4639	25.24	11.708
	2	0.7590	33.672	25.557
	3	0.88396	109.36p	96.669
Proposed (down)	1	0.4694	28.276	12.333
	2	0.7703	40.750	31.389
	3	0.8996	112.22	100.95

6.5 CONCLUSION

Multi-valued logic provides the advantages of enhanced processing capability and reduced power overhead due to interconnect complexity in VLSI design. Thus this work has presented CNTFET-based D-flip-flop and counter cells using radix 3 by adopting rotational symmetry of shifting unary literals. For the realization of shifting circuits, the average of unary operators together with efficient voltage divider topology is adopted, which is providing better performance metrics. The D flip-flops are built using a combination of these shifting operators implemented in a master-slave architecture. The proposed D-flip-flop cells are deployed to implement ternary counter structures to confirm the suitability of suggested circuits in realizing multistage sequential blocks. The HSPICE simulations are conducted with the 32-nm CNTFET Stanford model to test the functionality of the proposed modules. The experimental findings reveal that in terms of power consumption and energy efficiency, the proposed flip-flop and counter designs perform better than other state-of-art alternatives. Thus, sequential block implementation that makes use of multi-valued logic architecture makes it easier to take advantage of energy efficiency benefits by boosting the digital system's processing power.

REFERENCES

1. Sanjeet Kumar Sinha and Saurabh Chaudhary. Comparative study of leakage power in CNTFET over MOSFET device. *J. Semicond.*, vol. 35, no. 11, p. 114002, 2014.
2. D. Miller, and M. Thornton, *Multiple valued logic: Concepts and representations*. San Rafael, CA: Morgan & Claypool Publishers, 2008.
3. E. Dubrova, V. Sreehari, and M. B. Srinivas. Multiple-valued logic in VLSI: challenges and opportunities. In *NORCHIP Conference*, Oslo, Norway, pp. 340–350, 1999.

4. G. Hills, "Understanding energy efficiency benefits of carbon nanotube field-effect transistors for digital VLSI," *IEEE Trans. Nanotechnol.*, vol. 17, no. 6, pp. 1259–1269, 2018.

5. G. S. Tulevski. "Toward high-performance digital logic technology with carbon nanotubes," *ACS Nano*, vol. 5, pp. 8730–8745, 2011.

6. P. Prakash, K. Mohana Sundaram, and M. Anto Bennet. "A review on carbon nanotube field effect transistors (cntfets) for ultra-low power applications," *Renew. Sust. Energ. Rev.*, vol. 89, pp. 194–203, 2018.

7. R. Katreepalli, and T. Haniotakis Power efficient synchronous counter design. *Comput. Electr. Eng.*, vol. 75, pp. 288–300, 2019.

8. S. Lin, Y.-B. Kim, and F. Lombardi, "CNTFET-based design of ternary logic gates and arithmetic circuits," *IEEE Trans. Nanotechnol.*, vol. 10, no. 2, pp. 217–225, 2011.

9. M. H. Moaiyeri, R. F. Mirzaee, K. Navi, and A. Momeni, "Design and analysis of a high-performance cnfet-based full adder", *Int. J. Electron.*, vol. 99, pp. 113–130, 2012.

10. B. Srinivasu, and K. Sridharan. Low-complexity multiternary digit multiplier design in cntfet technology. *IEEE Trans. Circuits Syst. II: Express Br.*, 63(8):753–757, 2016.

11. B. Srinivasu, and K. Sridharan, A synthesis methodology for ternary logic circuits in emerging device technologies, *IEEE Trans. Circ. Syst. I: Regul. Pap.*, vol. 64, pp. 2146–2159, 2017.

12. C. Vudadha, S.P. Parlapalli, and M. Srinivas, Energy efficient design of cnfet-based multi-digit ternary adders, *Microelectron. J.*, vol. 75, pp. 75–86, 2018.

13. F. Zahoor, T. Z. A. Zulkifli, F. A. Khanday and S. A. Z. Murad, "Carbon nanotube and resistive random access memory based unbalanced ternary logic gates and basic arithmetic circuits", *IEEE Access*, vol. 8, pp. 104701–104717, 2020.

14. S. Tabrizchi, A. Panahi, F. Sharifi, K. Navi, and N. Bagherzadeh. Method for designing ternary adder cells based on cnfets. *IET Circuits, Devices Syst.*, vol. 11, no. 5, pp. 465–470, 2017.

15. K. Rahbari, and S. A. Hosseini, "Novel ternary D-flip-flap-flop and counter based on successor and predecessor in nanotechnology," *AEU - Int. J. Electron. Commun.*, vol. 109, pp. 107–120, 2019.

16. M. H. Moaiyeri, M. Nasiri, and N. Khastoo, "An efficient ternary serial adder based on carbon nanotube FETs," *Eng. Sci. Technol., Int. J.*, vol. 19, no. 1, pp. 271–278, 2016.

17. A. Thakur, and R. Mehra, Power and speed efficient ripple counter design using 45 nm technology, in: *IEEE Int. Conf. on Power Electronics, Intelligent Control and Energy Systems*, Delhi, 2016, pp. 01–04.

18. T. Sharma, and L. Kumre, "Design of unbalanced ternary counters using shifting literals based D-Flip-Flops in carbon nanotube technology," *Comput. Elect. Eng.*, vol. 93, 2021, Art. no. 107249.

19. J. Deng, and H. P. Wong, A compact SPICE model for carbon-nanotube field-effect transistors including nonidealities and its application Part I: model of the intrinsic channel region, *IEEE Trans. Electron. Dev.*, vol. 54, pp. 3186–3194, 2007.

20. J. Deng, and H. P. Wong, A compact SPICE model for carbon-nanotube field-effect transistors including non-idealities and its application part II: full device model and circuit performance benchmarking, *Electron Devices, IEEE Trans.*, vol. 54, pp. 3195–3205, 2007.

Chapter 7

Novel radiation-hardened low-power 12 transistors SRAM cell for aerospace application

Vancha Sharath Reddy, Arjun Singh Yadav and Soumya Sengupta
NIT Rourkela Odisha, India

7.1 INTRODUCTION

Space applications such as satellites and space stations are exposed to heavy radiation (high-energy particles), which interferes with the semiconductor-based electronic circuitry. The interaction of such radiation with semiconductors will have negative effects on circuit performances. The life requirement of a satellite is generally 10 years; therefore, they are required to operate reliably for long periods of time in harsh radiation environment without being serviced. Therefore, the need arises to make the circuits tolerant to radiation; this process is known as radiation hardening. Radiation hardening is achieved by shielding and adding redundancy to the circuits [1]. Radiation hardening by design (RHBD) refers to a method that makes the circuit tolerant to radiation by incorporating a mechanism into the circuit. Hence, a better choice to mitigate soft error is to design a radiation-hardened cell. Radiation impacts on CMOS circuits may be roughly classified into three types: total ionizing dose (TID), single-event effect (SEE), and displacement damage (DD) [2]. Total ionizing dose is when high-energy particles produced by radiation enter electronic devices and produce electron-hole pairs. Holes, being less mobile than electrons, get trapped in gate oxides, and the accumulating positive charge degrades the device's performance. Single-event effects take place when, in a split second, high-energy particles impact the nodes of a device, changing the voltages in the device as well as disrupting data in memory and flip-flops. Displacement damage is whereby high energy particles, when impacting a device, collide with a silicon atom in the crystal lattice and displace it from its position, which increases the leakage current and consequently degrades the performance of the device.

The main sources of radiation in space are Van Allen belts, where the magnetic fields trap the radiation from solar flares and cosmic rays [3]. All these radiation sources contain highly energized charged particles that include alpha, protons, and high-energy neutrons (>1 MeV). When these particles strike a logic circuit like SRAM, electron and hole pairs get generated by impact ionization as the particle loses energy. Electron and hole

DOI: 10.1201/9781032670270-7

pairs in the semiconductor get separated due to the presence of an electric field and subsequently accumulate at the sensitive node. Once accumulated, these charges create both positive and negative transient voltage pulses. When the amplitude of these pulses overcomes the switching threshold that affects the memory circuit at the drain end, the originally stored data may undergo changes, which results in an SEU [4]. This also sometimes is referred to as soft error, because the data altered by a radiation particle–caused SUE get replaced with the appropriate data in the next write operation. Thus, it results in a system malfunction but does not cause permanent damage [5]. Single-event effects (SEEs) are the major threats faced by SRAM cells in memory circuits. Memory circuits are further prone to an SEE, which can be one of two types: destructive and nondestructive [6].

Nondestructive SEEs change the node voltages of semiconductor device but do not lead to functional failure of device, whereas destructive SEEs change the state of the circuit that leads to the functional failure. Single-event upsets (SEUs) cause the node voltage of an SRAM cell or a latch to change, which causes the bit-flip in the circuit, thereby causing a soft error. An SEU occurs when a particle strike happens at a storage node. These particle strikes occur at the nodes of the Static RAM cell, which are sensitive to radiation. A sensitive node refers to a node that is connected to an OFF NMOS/PMOS transistor at the drain terminal. This node will act as a collection point of charges due to particle strike. As the charge increases at the node, the voltage value shifts, which further shifts the bit value stored in the circuit [7]. In a semiconductor device, the sensitive region is the strong reverse-biased diffusion region where transient current induced in the device passes from N-channel diffusion region to the P-channel diffusion region [8]. Hence, a positive transient pulse gets induced when the radiation particle strikes the PMOS transistor drain, and a negative transient pulse gets formed when the radiation particle strikes the NMOS transistor drain.

Suppose that the drain of a PMOS is at '0' and is stuck by radiation, and a positive transient pulse is generated, thereby producing a 0 ➔ 1 SEU. In case of a NMOS drain, a negative transient pulse is generated, thereby producing a 1 ➔ 0 SEU. Now suppose that the drain of a PMOS is at '1' and is struck by radiation; it produces a 1 ➔ 1 glitch, and if the drain of a NMOS is at '0', it produces a 0 ➔ 0 glitch. Hence, the latter case does not affect the stored data.

Several studies have been conducted that aim at preventing or mitigating the deleterious effects of radiation on integrated circuits. Gaul and colleagues [9] described the various effects of radiation on CMOS circuits, as well as the radiation sources. They also described different methodologies used to make a circuits radiation hardened. Pal and coworkers [10] proposed a radiation-hardened 10-transistor SRAM cell capable of recovering from SEUs with a better-read stability and critical charge in comparison to other cells that were considered in the study. The same group in another study [11] proposed a 14-transistor radiation-hardened SRAM cell that can provide recovery from SEU at every sensitive node as well as multiple-node upset at the node pair. The decoupling read technique improved the read stability of the cell. The strategy used in this circuit for tolerance of multiple-node upset is employed in the Proposed12T cell in this chapter. Saun and Kumar [12] described the characterization methods for determining the read and write margins of an SRAM cell. They also provided a comparative analysis of performance parameters of a 6T SRAM cell at various PTM technology nodes. Amusan and colleagues [13] described the effects of charge sharing leading to single- and multiple-node upset of an SRAM cell. They presented a solution to avoid charge sharing between nodes by maintaining the minimum spacing between them.

Several soft error–immune radiation-hardened memory cells also have been proposed. Authors in [14] have proposed a soft error–immune cell QUATRO-10T, which uses negative feedback but can only provide SEU recovery induced at the '1' storage node and is unable to provide recovery from an SEU induced at the '0' storage node. QUATRO-10T also possesses poor ability to write in the cell. A modified version of quadruple QUATRO-10T, known as WE-QUATRO, or QUATRO-12T [15], possesses an enhanced write ability but is still incapable of SEU recovery from a 0 storage node. Furthermore, it possesses poor stability during read mode and higher hold power. Another SRAM cell, QUCCE-10T [16], shows a longer write delay and is not capable of recovering a 0 → 1 after an SEU provided that sufficient charge gets stored at the storage node storing 0. QUCCE-12T suffers from poor stability to read, and hold power consumption is high, similar to WE-QUATRO.

Another cell, RHMD-10T [10], provides tolerance from SEU caused at all the sensitive nodes and also provides SEMNUs recovery, but it has a high probability of write operation failure. The conventional 10T SRAM cell (RHMD10T) was built to recover from SEUs at its sensitive nodes. A highly reliable radiation-hardened Static RAM cell (RHM-12T) with 12 transistors has been proposed [8] that provides sufficient protection from soft errors at its sensitive nodes not only from SNUs but also from multi-node upsets (MNUs) induced by the sharing of charge on the fixed nodes, with better reliability. RHM-12T provides lower static power and lower access time. It also has higher SEU tolerance capability. The only drawback is that of a lower read SNM, which may turn out to be not effective for some application Further, modeling and radiation hardening effects analysed in MOS device and digital microelectronics [17, 18]

In this chapter, we propose a novel radiation-hardened low-power Proposed12T SRAM cell that is capable of fully recovering from SEUs, at both polarities, produced at any sensitive nodes. The Proposed12T cell provides recovery not only from SEUs at sensitive nodes but also from MNUs at the storage pair (S0-S1), which is achieved by isolating the storage nodes (Q-QB) from S0-S1 using two NMOS transistors in the pull-up path. This isolation strategy prevents the fault caused by MNUs.

The chapter is organized as follows: Section 7.2 describes the fundamental structure and working operation of the Proposed12T cell along with the SEU node recovery analysis. Section 7.3 provides the simulation setup along with the detailed comparison of the characterization parameters in terms of read access time (T_{RA}), write access time (T_{WA}), read static noise margin (RSNM), write static noise margin (WSNM), hold static noise margin (HSNM), leakage power consumption (P_{Hold}), critical charge (Q_C), total area (A_{total}), and SEU occurrence probability. Finally, the concluding part of the summary is provided in Section 7.4.

7.2 THE PROPOSED12T CELL AND ITS OPERATIONS

7.2.1 Cell structure and working operation

The Proposed12T Static RAM Cell schematic diagram is presented in Figure 7.1, and the respective layout of the Proposed12T is shown in Figure 7.2. The Proposed12T SRAM cell has six PMOS and six NMOS transistors. In the Proposed12T, a word line signal controls NMOS N3 and

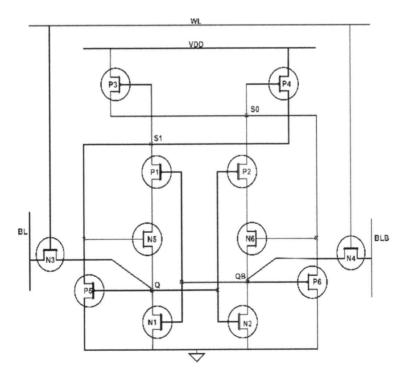

Figure 7.1 Proposed12T SRAM cell schematic diagram.

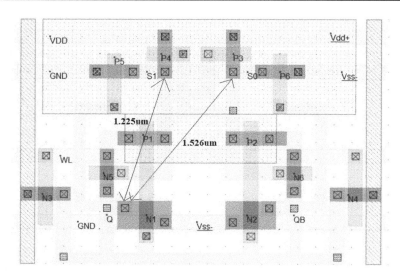

Figure 7.2 Compact cell layout of the Proposed12T cell.

N4, where the charge storage nodes Q and QB are connected with the BL and complementary BL. Furthermore, Proposed12T contains two internal nodes in the form of S0 and S1. If we consider a case where the Proposed12T is storing '1', it indicates that storage nodes Q is storing '1' and QB is storing '0' value and the internal cell nodes S0 stores '0' value and S1 storing '1' value. In the Proposed12T cell, two NMOS transistors in the pull-up path are added in order to avoid bit-flip caused by the (S0-S1) SEMNU.

This topology also has same number of nodes, i.e., S0, QB, S1, and Q. The data at nodes QB and Q are held stable because of the cross-coupled latch between P2-N2 and P1-N1 transistors. However, because of the two NMOS transistors in the pull-up path, they cannot pass a full logic '1' value. This affects the voltage swing at nodes Q and QB, which reduces the SNM of the cell.

The nodes S0 and S1 are also held stable because of the cross-coupled connection between PMOS P4 and P3.

1. In hold state, access transistors are tuned in OFF condition by setting the word line to GND, and the bit lines are pulled up to VDD by charging. Consider a logic '1' write operation: the node Q has been charged to '1' through access transistors. Node Q is driving the PMOS transistors P2 and P5, and NMOS transistor N2. As a result, P2 and P5 are switched OFF and N1 is switched ON. The node QB is discharged to '0' through transistor N2, QB is driving PMOS transistors P1 and P6, and NMOS transistor N1. Hence, PMOS transistors P6 and P1 are turned ON and NMOS N1 is turned OFF. P6 pulls down

the node S0 to '0'. Node S0 is driving the gates of PMOS transistor P4, and NMOS transistor N6. P4 pulls up the node S1 to '1', as S1 is driving NMOS transistor N5 and PMOS transistor P3. N5 turns ON and P3 turns OFF. Therefore, we have a conducting path from transistors P4, P1, and N5, which holds the logic '1' at node QB stable. Similarly, the pull-down transistor N2 holds the logic '0' at node QB stable.

2. In order to perform the operation during read mode, both bit lines are pre-charged to supply voltage. The active mode is set for the access NMOS transistors N3 and N4 by setting WL to VDD. As a result, the bit line across QB gets discharged through N4 along with N2. As N1 remains in OFF state, BL stays at VDD. When the potential difference between the bit lines reaches 50 mV, the stored data performed during the read cycle is sensed by a sense amplifier [4]. For reliable operation during read mode, cell ratio (CR), which is defined as the ratio of NMOS transistor N1 and N3 or that of N2 and N4, is set to 2.

3. To perform write operation, access transistors N3 and N4 are turned ON. The BL is kept at '0' and BLB is kept at '1'. Initially, charge-storing nodes Q and QB are kept at '1' and '0', respectively. The pulling-down of node Q toward ground terminal (GND) and pulling-up of node QB toward the power supply terminal (VDD) are started by both bit lines BL and BLB. Hence, node Q voltage is pulled down to the ground terminal (GND) by bit line through access transistor N3, and as a result, node Q will turn OFF N2 because it reaches below the threshold voltage line of the N2 transistor and turns ON P5. On the other hand, this boosts up the charging of QB, and QB is pulled up by the complementary bit line (BLB) through N4. Thus, node QB turns N1 ON and P6 is turned OFF. S1 gets discharged to GND through P5 (as P5 is ON), which turns ON P3 and P2. P6 being turned OFF and P3 being turned ON enables S0 to get charged up. Hence, both P1 and P4 become OFF. Thus '0' is written at node Q and '1' is written at node QB. In this way, the write operation is performed that flips the states of the internal node to the desired value.

7.2.2 SEU recovery analysis

The Proposed12T cell provides recovery at all sensitive nodes. The sensitivity of a node is defined by the region that falls within the vicinity of a diffusion-based OFF transistor drain region, which is reverse-biased [8]. When the bit cell has a logic '0', the nodes S1 and QB are sensitive to SEU because they are connected to the drain terminal of an OFF transistor. Similarly, when SRAM bit cell has a logic '1' value, the charge-storing nodes Q and S0 are linked to drain terminals of OFF transistors. Here, initially, the cases assumed are Q = 1, QB = 0, S0 = 0, and S1 = 1, and Q and S0 are sensitive to SEU at logic 1. We can conclude that in either of the cases the Proposed12T

SRAM cell is sensitive to SEUs at two of its nodes. The SEU recovery analysis of the two sensitive nodes Q and S0 and for node-pair S0-S1 is given below:

1. Node Recovery at Q (1→0): The charge-storing node Q is storing logic '1' when a negative transient SEU causes a 1→0 transition. P2 and P5 are turned ON and NMOS N2 is turned OFF. As N2 becomes OFF, QB is driven to high impedance state. Even though P5 is ON, it cannot pull down S1 to '0' because of the superior strength of P4. The accumulated negative charge at node Q is pulled up to VDD through PMOS transistors P1, P4, and NMOS N5. Thus, Q is able to recover from SEU.
2. Node Recovery at S0 (0 →1): The node S0 is storing a logic '0' when a positive transient SEU causes a 0→1 transition; P4 becomes OFF and N6 becomes ON temporarily. As a result, S1 is set to a high impedance condition. After a certain time, positive charges accumulated are drained into the ground through P6 pull-down. Thus, S0 is able to recover from SEU.
3. SEMNU Recovery at Nodes S0-S1: The cell is holding a bit '1', so the node voltages are Q = 1, QB = 0, S1 = 1, and S0 = 0. When a SEMNU occurs on the nodes S0-S1, the node S0 is affected by positive transient and S1 is affected by negative transient. As a result, S0 will have a 0→1 transition and S1 will have a 1→0 transition. S0 is driving P4 and N6, and S1 is driving P3 and N5. So, N5 and P4 are switched OFF while N6 and P3 are switched ON. N5 does not allow the fault in S1 to propagate to Q, also P2 is OFF so it does not allow the fault in S0 to propagate to QB. After some time, the negative charge at node S0 is discharged via P6 pull-down path, restoring the initial voltage of node S1. Thus, SEMNU at nodes S0-S1 does not cause the bit-flip in the cell.

7.3 SIMULATION SETUP AND RESULTS

The simulation is carried out using 65-nm CMOS PTM technology using HSpice tool, and the supply voltage is 1V with cell ratio and pull-up ratio as 2 and 1, respectively. The Proposed12T SRAM cell is compared with several radiation-hardened SRAM cells such as the RHMD-10T, RHM-12T, QUATRO-12T, and QUCCE-12T. The sizing of the cells chosen is compared i the respective paper for comparative analysis purpose [5–9].

7.3.1 Read delay comparison

During a read operation, both BLB and BL are pre-charged to VDD. When WL is made active to read the data, the time taken for BL (read '0') or

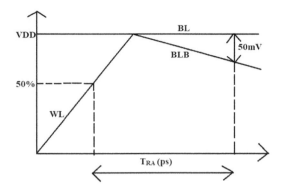

Figure 7.3 Read access time definition.

BLB (read '1') to drop by 50 mV from 50% of WL is called read access time [8]. Figure 7.3 illustrates the access time during read operation. Read delay or access time during read mode (T_{RA}) depends on bit line capacitance and read current. Hence, the access time during read mode of the cell is mainly dependent on read current passing through the pull-down NMOS and through the access transistors (N3 and N4), which in turn is dependent on the body current and cell ratio (CR) [19]. The largest body effect offered by the pull-down path of stacked NMOS representation offers lowest read current. The higher the CR of both read paths, the lower the read access time (T_{RA}). Extra read path results in shorter T_{RA}. Because of the presence of voltage divide effect between the pass transistors and NMOS pull-down transistors, the charge-storing node QB that stores '0' rises during the read operation. As a result, the driving strength of the pass transistors also gets reduced due to body effect. Due to the reduced driving capabilities of pull-down transistors (N1 and N2), it increases cell access time during read mode (T_{RA}). Hence, the Proposed12T cell has the highest T_{RA} in comparison to the other selected cells. As QUATRO-12T and QUCCE-12T have two pass transistors linked to a bit line, they have higher BL capacitance than RHMD-10T. The Proposed12T cell has a marginal penalty of read delay and is 7.67%/63.14%/28.87%/36.36% of RHMD-10T/RHM-12T/QUATRO-12T/QUCCE-12T SRAM cells.

7.3.2 Static noise margin comparison

The static noise margin (SNM) helps test the stable state of a static RAM cell [9]. SNM is defined as the smallest amount of noise voltage capable of changing the value of a bit stored in an SRAM cell. The SNM can be calculated by considering the maximum length of a side of the largest square that fits in the VTC Butterfly curve of Q and QB [20, 21] as can be seen in Figures 7.5 to 7.7, respectively. An SRAM cell has three types of SNM:

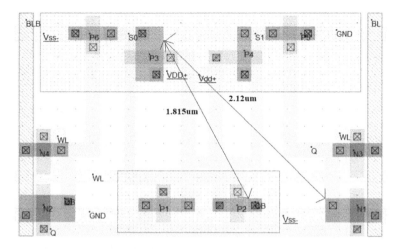

Figure 7.4 Compact cell layout of RHMD-10T.

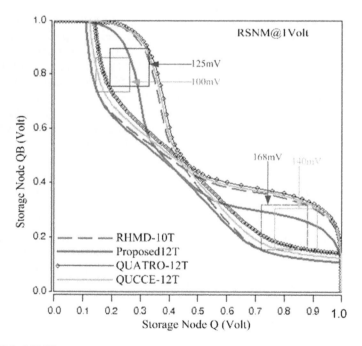

Figure 7.5 RSNM comparison.

read static noise margin (RSNM) measured during the read operation; write margin (WM) measured during the write operation; and hold static noise margin (HSNM) measured while the cell is in the hold state. It is observed in conventional 6T SRAM cells (based on direct read mechanism), the value

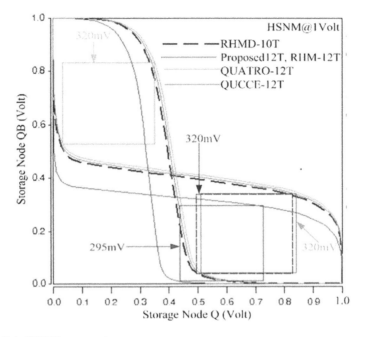

Figure 7.6 HSNM comparison.

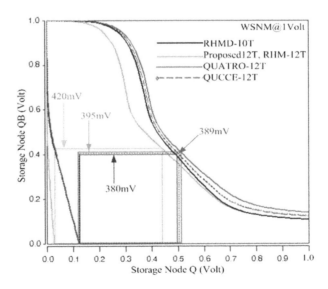

Figure 7.7 WSNM comparison.

of RSNM varies between 10% and 20% of V_{DD}, while WSNM/HSNM is about 50% of V_{DD}. Hence, RSNM is the most critical parameter in SRAM design.

Read stability of all the cells is measured by the RSNM of a memory cell. The readability (RSNM) of an SRAM cell mainly depends on the cell ratio (CR) value and read path [5]. During read analysis, when the word line signal is enabled, the bit lines (BL and BLB) are activated and the storage nodes gets accessed from the bit lines (BL and BLB). In this chapter, the Proposed12T SRAM cell has been compared with RHMD-10T, QUATRO-12T and QUCCE-12T (Figure 7.5). The higher the cell ratio of the SRAM cell, the lower the voltage at the 0-storing node QB, and hence it provides a good RSNM. Since the stacking effect is more in the Proposed12T SRAM cell (a series of PMOS and NMOS transistors are stacked), a higher voltage gets generated at the node QB that stores 0, thereby reducing the RSNM. Hence, RSNM of the Proposed12T SRAM cell is slightly lower in comparison to the other cells mentioned here. Thus, the Proposed12T cell has shown a marginal penalty of read margin compared to the other cells.

Similarly, in hold mode, the HSNM has been measured by deactivating the bit lines, and as a result, the bit lines cannot access the charge-storing nodes. The hold SNM of the Proposed12T SRAM cell is compared with RHMD-10T, RHM-12T, QUATRO-12T, and QUCCE-12T as seen in Figure 7.6. The Proposed12T cell has similar HSNM as that of RHM-12T cell. It is observed that the Proposed12T cell has shown 7.81% lower HSNM compared to RHMD-10T/QUATRO-12T and QUCCE-12T. Overall, the Proposed12T cell possesses the lowest HSNM compared to other cells.

In write mode, only the access transistor across BLB and QB is activated. The write SNM comparisons for every memory cell mentioned in the chapter are shown in Figure 7.7. Due to similar transistor stacking, the Proposed12T cell has the same WSNM as that of RHM-12T cell. The WSNM of the Proposed12T cell is 110.52%/106.32%/107.96% of RHMD-10T/QUATRO-12T/QUCCE-12T. Overall, the Proposed12T cell along with RHM-12T cell possess the highest WSNM as can be seen in Table 7.1 compared to RHMD-10T, QUATRO-12T, and QUCCE-12T.

7.3.3 Write delay comparison

During the write operation, the amount of time taken between 50% of WL to the intersection point of both bit lines is called the write access time [8]. The cell access time during the write mode for the Proposed12T cell has been compared with that of RHMD-10T, RHM-12T, QUATRO-12T, and QUCCE-12T. It is observed from Table 7.1 that RHMD-10T has offered the highest T_{WA} and RHM-12T has offered the lowest T_{WA} with respect to the other comparison cells. Due to the stacking effect, the Proposed12T cell has a higher T_{WA} compared to the other comparison cell as it varies with marginal increase. RHM-12T has the lowest T_{WA} due to stacked PMOS

Table 7.1 Comparison among various radiation-hardened SRAM Cells at VDD = 1 V

Design metric	RHMD-10T [5]	RHM-12T [6]	QUATRO-12T [8]	QUCCE-12T [9]	Proposed-12T
TRA (ps)	100.3	66.20	83.80	79.2	108
TWA (ps)	23.64	5.50	17.90	18.24	26.5
RSNM (mV)	168	8.5	125	140	100
HSNM (mV)	320	295	320	320	295
WSNM (mV)	380	420	395	389	420
QC (fC)	>39.45	>50	10.34	11	43
PHold (nW)	3.713	23.492	6.02	6.18	3.612
Relative Area	0.86	1	0.92	0.903	1
Sensitive Nodes	3	3	4	4	3
PS	0.034	0.0494	0.035	0.033	0.048
Relative CNPDA	2.065	0.119	0.381	0.461	1

structure. As two additional pass transistors connect to the internal nodes of the two inverter pairs that assist in the write operation, the write operation of QUATRO-12T and QUCCE-12T gets enhanced [5, 14]. As a result, QUATRO-12T and QUCCE-12T have a lower T_{WA} compared to RHMD-10T. RHM-12T has a lower write access time compared to the other memory cells due to the decreased driving capabilities at storage nodes Q and QB. The Proposed12T cell has shown the highest write access time than all other comparison cells.

7.3.4 Hold power comparison

The criteria that is essential while considering the hold power is that a smaller number of transistors lead to lower leakage consumption, presence of stacked transistor structure leads to lower P_{Hold} dissipation, stacked NMOS structure offering larger body effect to the pass transistors dissipates to even lower P_{Hold}, and power consumption can be reduced by scaling down the VDD. The two major sources of hold power (P_{Hold}) consumption are bit-line leakage and leakage in the inverter. The Proposed12T cell has low hold power, since the stacked transistors N5 and N6 are present in the cell. Hence, leakage power dissipation is considerably reduced. Since the radiation-hardened memory cell RHMD-10T contains the PMOS transistors (P5 and P6) in the form of weak pull-down path, the leakage power dissipation is smaller compared to most of the comparison cells. QUATRO-12T and QUCCE-12T, having four access transistors, including two other additional pass transistors, consume higher P_{Hold}, resulting in more leakage components in the bit lines. Hence, they have higher P_{Hold} value. The leakage power consumption of the Proposed12T cell is reduced up to 97.27%/60%/58.44% compared to RHMD-10T/QUATRO-12T/QUCCE-12T cells.

7.3.5 Soft-error recovery simulations and comparison

According to the literature, the following models can be employed for estimation of the SEU tolerance or to characterize Q_{crit}:

   i. Simplified model by Roche
   ii. Current model by Freeman
   iii. Diffusion collection model
   iv. Double Exponential Current Model

In this chapter, to perform the simulation for the impact of radiation on semiconductor device, double exponential current source has been considered [14, 16, 22, 23]:

$$I(t) = I_0 \times \left\{ e^{-t/t_\beta} - e^{-t/t_\gamma} \right\}$$

$$I_0 = Q_{coll}/(t_\beta - t_\gamma)$$

where I_0 is the current pulse amplitude or the peak current of the current pulse, Q_{coll} is the quantity of collected charge that gets generated by collision among high-energy particles or the injected charge, t_β is the collection time constant of a junction, and t_γ is initial ion-track establishing time constant. Critical charge (Q_{crit}) is defined as the lowest amount of charge accumulated at a sensitive node that can flip the original data stored in the memory cell. During evaluation, the Q_{crit} that contains the lowest value among all the sensitive nodes gives the effective Q_{crit} of the cell [16]. Here the estimated Q_{crit} is considered at 43fC. This estimated charge causes the stored data to flip temporarily when deposited at any single node (Q, S0). The current is fed to the sensitive nodes – that is, across Q and S0. After estimating the critical charge across those sensitive nodes, the lowest among them is considered the effective critical charge of the memory cell. The source polarity of the current is adjusted across the target sensitive node to produce a positive transient pulse at the drain terminal of OFF PMOS transistor and to generate a negative transient pulse at the drain terminal of OFF NMOS transistor. RHM-12T shows a higher level of soft-error tolerance and has the highest critical charge with respect to other comparison cells as can be seen in Table 7.1. The Proposed12T cell has shown 8.99% improvement in critical charge over the RHMD-10T cell.

7.3.6 Layout area and SEU probability comparison

In order to compare the area and get a detailed analysis, a layout has been designed. The layout of only RHMD-10T has been shown in Figure 7.4

for comparison. The relative area comparison of all the cells is shown in Table 7.1. Since large number of transistors are present, especially PMOS transistors, the relative area of the Proposed12T cell is larger. The layout analysis of the existing RHMD-10T has been compared with the layout of Proposed12T SRAM cell. A large number of transistors increases the layout area. The charge sharing between nodes at close proximity will cause SEMNU, so in order to avoid charge sharing between NMOS-PMOS, PMOS-PMOS, and NMOS-NMOS, a distance of 0.6 μm, 1.62 μm, and 2 μm should be maintained, respectively [8, 13]. The Proposed12T layout as shown in Figure 7.2 has been performed using four different types of metals. Metal 4 is used for connecting BL to the source terminal of N3 and BLB to source of N4, whereas metal 3 is used for connecting the word line to the gate terminal of N3 and N4. In the layout of Proposed12T SRAM cell, the distance between Q-S0 is 1.526 μm (35 × 25 lambda) and for QB-S0 and Q-S1 the distance is the same, namely 1.225 μm (35 lambda). All these node pairs have a spacing greater than the charge-sharing range between NMOS-PMOS, i.e., 0.6 μm. Therefore, the possibility of occurrence of SEMNU at these nodes is minimized.

In the layout of RHMD-10T SRAM cell in Figure 7.4, the distance between Q-S0 (PMOS–PMOS) is 2.12 μm and for QB-S0 (PMOS–PMOS) and Q-S1 it is 1.815 μm. The distance between Q-QB (PMOS–NMOS) is 0.63 μm. The minimum distance between these transistors is targeted at the sensitive nodes that change value or state due to radiation. Hence if minimum distance is maintained in the layout, the chances of SEMNU get reduced considerably. Further, both layouts are free from DRC and no error has been found. The area occupancy of the Proposed12T SRAM Cell is 100 × 70 lambda, whereas the total area occupancy of the RHMD-10T is 100 × 60 lambda.

The provided results are extracted from the layout at CMOS PDK 65-nm technology. It has been found from the layout that the parasitic capacitance of Proposed12T SRAM cell at nodes Q, QB, S0, and S1 are 2.44, 2.34, 1.29, and 1.41, respectively. Also, the BL/BLB parasitic capacitance is 0.13. In comparison, the parasitic capacitance of RHMD-10T cell at nodes Q, QB, S0, and S1 is 2.11, 1.86, 1.47, and 1.97, respectively. As a result, the parasitic capacitance for the proposed model is slightly higher for nodes Q and QB because of the longer path connected to a higher number of transistors in comparison to RHMD-10T. However, the proposed layout has been able to reduce the parasitic capacitance for nodes S0 and S1 because in the proposed model, nodes S0 and S1 are connected to only three transistors at the sensitive node, whereas in RHMD-10T cell, sensitive nodes S0 or S1 are connected to five transistors, making both area and parasitic capacitance greater. Furthermore, the parasitic capacitance values extracted from the layout for the proposed model have been used in the schematic for noise margin analysis.

The probability of occurrence (P_S) of an SEU can be estimated from the designed layout. It is defined as the probability of a static RAM cell that gets impacted by an SEU. It is given as:

$$P_S = \frac{A_S}{A_{total}} \tag{7.1}$$

where A_s is the area covered by the sensitive region and A_{total} is the overall area of the static RAM cell. The lower the probability occurrence value as shown in Equation (7.1), the lesser the probability of a node being affected by an SEU. The smaller number of sensitive nodes leads to lower P_S, and the large area overhead also leads to less P_S. Hence, it is important to have a slight area overhead and fewer sensitive nodes, whose drain terminal of the MOS transistor is sensitive, resulting in decreasing A_S of the cell. Here, the OFF NMOS and OFF PMOS are used for the estimation of the radiation-sensitive area. Since fewer sensitive nodes are being occupied by the Proposed12T cell, the area occupied by the sensitive nodes is also smaller. Hence, the Proposed12T cell has the least SEU occurrence probability compared to its comparison cells. QUATRO-12T and RHM-12T have the larger area overhead.

7.3.7 Enhanced quality metric for static RAM cell (CNPDA)

In today's technology, a cell that contains smaller area, enhanced SNM, soft-error tolerance, low power dissipation, and optimum delay is preferred. However, a certain trade-off takes place in improving the design metrics of the cell. Considering all these parameters, to access the complete performance of the cell, a new metric of design called critical charge-NM to power-delay-area (CNPDA) comes into play:

$$CNPDA = \frac{Q_{crit} \times RSNM \times HSNM \times WSNM}{P_{Hold} \times T_{WA} \times T_{RA} \times Area} \tag{7.2}$$

A cell with higher CNPDA derived from Equation (7.2) shown is much preferred for overall performance. From the preceding discussion, it is obvious that the Proposed12T cell has the highest CNPDA compared to RHMD-10T, RHM-12T, QUATRO-12T, and QUCCE-12T.

7.4 CONCLUSION

This chapter has presented the radiation-hardened Proposed12T SRAM cell that provides recovery from single-node as well as multi-node upset.

The design metric analysis has been done and has compared with RHMD-10T, RHM-12T, QUATRO-12T, and QUCCE-12T. In order to tolerate SEMNU at other node pairs, a layout design is made such that the charge sharing between the nodes can be prevented by appropriate spacing; this design also has been compared with existing work. The Proposed12T cell has shown better performance against MNUs at the S0-S1 pair of nodes. The Proposed12T SRAM cell has shown improvement in critical charge and consumption of lower hold power in comparison to other cells. Additionally, the number of radiation-sensitive nodes of the Proposed12T cell has decreased, and as a result, the SEU occurrence probability has also slightly decreased. Overall, the performance metric CNPDA has also shown that the Proposed12T cell provides much better performance compared to RHM-12T, QUATRO-12T, and QUCCE-12T. Hence, the Proposed12T cell is a good alternative for application on space devices.

REFERENCES

[1] Sebestyen, G., et al., "Radiation Hardening, Reliability and Redundancy", in *Low Earth Orbit Satellite Design*, 2018, Springer. pp. 203–208.

[2] LaBel, K.A., et al., "Compendium of Single Event Effects, Total Ionizing Dose, and Displacement Damage for Candidate Spacecraft Electronics for NASA", in *2014 IEEE Radiation Effects Data Workshop (REDW)*. 2014. IEEE.

[3] Li, W., and M. Hudson, "Earth's Van Allen radiation belts: From discovery to the Van Allen Probes era." *J. Geophys. Res. Space Phys.*, vol. 124, no. 11, pp. 8319–8351, 2019.

[4] S. Pal, S. Bose, W. H. Ki, and A. Islam, "Half-select- free low-power dynamic loop-cutting write assist SRAM cell for space applications," *IEEE Trans. Electron Devices*, vol. 67, no. 1, pp. 80–89, 2020.

[5] S. Pal, D. D. Sri, W.-H. Ki, and A. Islam, "Soft-error resilient read decoupled SRAM with multinode upset recovery for space applications," *IEEE Trans. Electron Devices*, vol. 68, no. 5, pp. 2246–2254, 2021, doi: 10.1109/TED.2021.3061642

[6] S. Pearton, et al., "Opportunities in single event effects in radiation-exposed SiC and GaN power electronics", *ECS J. Solid State Sci. Technol.*, vol. 10, no. 7, pp. 075004, 2021.

[7] F. Wang, and V. D. Agrawal, "Single Event Upset: An Embedded Tutorial," 21st International Conference on VLSI Design (VLSID 2008), *IEEE Computer Society*, 2008.

[8] J. Gao, L. Xiao, and Z. Mao, "Novel low – power and highly reliable radiation hardened memory cell for 65nm CMOS technology," *IEEE Trans. Circuits Syst. I, Reg. Papers*, vol. 61, no. 7, pp. 1994–2001, 2014.

[9] S. J. Gaul, S. H. Voldman, and W. H. Morris, "Integrated circuit design for radiation environments", *General & Introductory Electrical & Electronics Engineering, Circuit Theory & Design*. 2019. 392.

[10] Pal, S., Sri, D. D., Ki, W. H., and Islam, A., "Highly stable low power radiation hardened memory- by-design SRAM for space applications", *IEEE Trans. Circuits Syst. II: Express Br.*, vol. 68, no. 6, pp. 2147–2151, 2020.

[11] S. Pal, S. Mohapatra, W. H. Ki, and A. Islam, "Soft-error-aware read-decoupled SRAM with MultiNode recovery for aerospace applications", *IEEE Trans. Circuits Syst. II: Express Br.*, 68(10), 3336–3340, 2021.

[12] S. Saun and H. Kumar, (2019, October), "Design and performance analysis of 6T SRAM cell on different CMOS technologies with stability characterization", In *IOP conference series: materials science and engineering* (Vol. 561, No. 1, p. 012093). IOP Publishing.

[13] O. A. Amusan, A. F. Witulski, L. W. Massengill, B. L. Bhuva, P. R. Fleming, M. L. Alles, and R. D. Schrimpf, "Charge collection and charge sharing in a 130 nm CMOS technology," *IEEE Trans. Nucl. Sci.*, vol. 53, no. 6, pp. 3253–3258, 2006.

[14] S. M. Jahinuzzaman, D. J. Rennie, and M. Sachdev, "A soft error tolerant 10T SRAM bit-cell with differential read capability," *IEEE Trans. Nucl. Sci.*, vol. 56, no. 6, pp. 3768–3773, 2009.

[15] L. D. T. Dang, J. S. Kim, and I. J. Chang, "We-quatro: Radiation hardened SRAM cell with parametric process variation tolerance," *IEEE Trans. Nucl. Sci.*, vol. 64, no. 9, pp. 2489–2496, 2017.

[16] J. Jiang, Y. Xu, W. Zhu, J. Xiao, and S. Zou, "Quadruple cross-coupled latch-based 10T and 12T SRAM Bit-cell designs for highly reliable terrestrial applications," *IEEE Trans. Circuits Syst. I, Reg. Papers*, vol. 66, no. 3, pp. 967–977, 2019.

[17] P. E. Dodd, and L. W. Massengill, "Basic mechanisms and modeling of single-event upset in digital microelectronics," *IEEE Trans. Nucl. Sci.*, vol. 50, no. 3, pp. 583–602, 2003.

[18] H. L. Hughes, and J. M. Benedetto, "Radiation effects and hardening of MOS technology devices and circuits," *IEEE Trans. Nucl. Sci.*, vol. 50, no. 3, pp. 500–521, 2003.

[19] C.-Y. Hsieh et al., "Independently-controlled-gate finFET Schmitt trigger sub-threshold SRAMS," *IEEE Trans. VLSI Syst.*, vol. 20, no. 7, pp. 1201–1209, 2012.

[20] S. Pal, S. Bose, W. H. Ki, and A. Islam, "Characterization of half-select free write assist 9T SRAM cell," *IEEE Trans. Electron Devices*, vol. 66, no. 11, pp. 4745–4752, 2019.

[21] R. C. Baumann, "Soft errors in advanced semiconductor devices-Part I: the three radiation sources," *IEEE Trans. Device Mater. Rel.*, vol. 1, no. 1, pp. 17–22, 2001.

[22] C. Qi, L. Xiao, T. Wang, J. Li, and L. Li, "A highly reliable memory cell design combined with layout-level approach to tolerant single-event upsets," *IEEE Trans. Device Mater. Rel.*, vol. 16, no. 3, pp. 388–395, 2016.

[23] M. Pown, and B. Lakshmi, *Investigation of Radiation Hardened TFET SRAM Cell for Mitigation of Single Event Upset*, School of Electronics Engineering, Vellore Institute of Technology, Chennai, India, 2020.

Chapter 8

Nanoscale CMOS static random access memory (SRAM) design

Trends and challenges

Sunanda Ambulkar

National Institute of Technology, Puducherry, India

Jitendra Kumar Mishra

NXPIndia Private limited, Noida, India

8.1 OVERVIEW OF TYPICAL 6T SRAM CELL

The typical SRAM cell is made up of four NMOS and two PMOS transistors, hence it is known as 6T SRAM cell [1, 2]. Figure 8.1 displays the basic design of the 6T SRAM cell. It comprises two cross-coupled CMOS inverter pairs, called latch, and two NMOS transistors, called access transistors. The transistors NM_1, NM_2, PM_1, and PM_2 form the internal latch while transistors NM_3 and NM_4 acts as access transistor for accessing the data from the cell and in the cell [3]. In the CMOS inverter the upper device, i.e., PMOS, is called a pull-up device or network, while the lower device, i.e., NMOS, is called a pull-down device or network. This structure has some inputs and output lines for accessing external data. A bit line (BL) and bit line bar (BLB) are complementary to each other, which works as data line input and output lines, which are connected column-wise while a word line (WL) is connected row-wise to enable the SRAM cell in the particular row to access the data. The typical 6T SRAM cell has two storing nodes, Q and Qb [4–6]. The SRAM cell works in different operating modes, i.e., write, read, and hold, as discussed in the following sections.

8.2 READ OPERATION

The behavior of a typical 6T SRAM cell for the read operation has been displayed in Figure 8.2(a). In this illustration, node Q starts with a logic '0' and node Qb starts with a logic '1' (logic '0' represents 0V and logic '1' represents maximum voltage or V_{DD}). As shown in Figure 8.2(a), the dotted-line transistors, PM_1 and NM_2, are in the OFF state whereas NM_1 and PM_2 are in the ON state. Because pre-charging the bit lines is required during the read operation, both bit lines are pre-charged to V_{DD} before the read operation begins [4]. The read operation has been initiated by turning

DOI: 10.1201/9781032670270-8

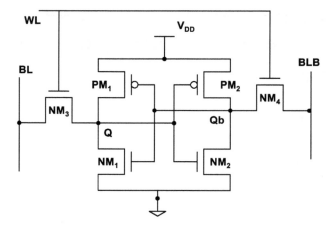

Figure 8.1 Typical 6T SRAM cell [3].

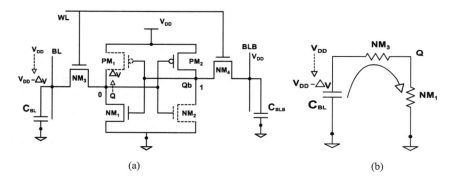

Figure 8.2 (a) Read operation of typical 6T SRAM cell. (b) Linear model of an equivalent resistor of transistors.

ON the access transistor via the activation of the WL. When the WL reaches logic '1', transistor NM1 enters the triode region, while transistor NM3 enters the saturation region. The current associated with NM1 and NM3 is proportional to the voltage at node Q [6]. Thus, it behaves like a resistor and forms a voltage divider circuit as shown in Figure 8.2(b). Because of this voltage divider circuit, the voltage at node Q exceeds ΔV from the initial value. So, in order to maintain the destruction-free read operation, the voltage ΔV should not exceed the threshold voltage of the NM_2 transistor. Figure 8.2(b) shows the linear resistive model for the discharge path of the bit line. The bit line capacitor C_{BL} is pre-charged to V_{DD}, and it starts to discharge through transistors NM_3 and NM_1 by the activation of WL [7]; this configuration causes the voltage drop at the bit line. Since the gate and source of the transistor NM_4 are at the same potential, i.e., $V_{GS4} = 0$ V, the

C_{BLB} cannot discharge and remains at V_{DD}. Hence a voltage difference of ΔV generated between BL and BLB, which is very small in value, is amplified by a sense amplifier to get a recognizable logic level. By lowering the resistance in the discharge path, the reading speed can be increased. This can only be possible by increasing the width of the transistors in this path, but at the same time this approach limits the density of the SRAM.

The read stability of the SRAM cell is also impacted by voltage ΔV during the read operation, so to make sure that the read operation is destruction free, the voltage ΔV must be controlled by the ratio of a resistor of NM3 and NM1. The cell ratio (β) parameter ensures the stability of the read operation. It is the ratio of W/L of the pull-down transistor to W/L of the access transistor [7]. The β value can be calculated by equating the current equations of transistors NM3 and NM1 in a specific operating region, which is given as:

$$I_{NM1} = I_{NM3} \tag{8.1}$$

$$\frac{\left(\frac{W}{L}\right)_{NM1}}{\left(\frac{W}{L}\right)_{NM3}} = \frac{\left(V_{dd} - 2V_{tn}\right)^2}{2\left(V_{dd} - 1.5V_{tn}\right)V_{tn}}$$

$$\text{Cell ratio}(\beta) = \frac{\left(\frac{W}{L}\right)_{NM1}}{\left(\frac{W}{L}\right)_{NM3}}$$

where $\left(\frac{W}{L}\right)_{NM1}$ is the aspect ratio of NM1 (pull-down) transistor, while $\left(\frac{W}{L}\right)_{NM3}$ is the aspect ratio of access transistor NM3 as given in Figure 8.1(a). A higher value of cell ratio (β) reduces the ΔV, which results in better read stability. The following section will cover the concept of stability in SRAM cell.

8.3 WRITE OPERATION

The behavior of typical 6T SRAM for data write operation is illustrated in Figure 8.3(a). It is initially assumed that nodes Q and Qb have logic '0' and '1', respectively. Thus, transistors NM2 and PM1 are turned OFF, while transistors NM1 and PM2 are turned ON. In the writing operation, the data on the cell must be changed [3]. As shown in Figure 8.3(a), all of the input conditions have been provided. The write operation is performed

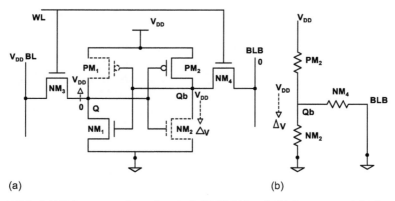

(a)   (b)

Figure 8.3 (a) Write operation of typical 6T SRAM cell. (b) Linear model of equivalent resistor of transistors.

by lowering the bit line voltage associated with node Q, i.e., BL, to a very
low voltage, i.e., 0V. Before the WL is activated, the transistors PM2 and
NM4 form a voltage divider circuit as demonstrated in Figure 8.3(b), which
influences the SRAM cell's writing ability. The pull-up ratio (γ) indicates
the write ability of an SRAM cell. The pull-up ration (γ) is calculated as the
ratio of the W/L of the pull-up transistor to the W/L of the access transistor.
It is given as follows:

$$I_{PM2} = I_{NM4} \tag{8.2}$$

$$\frac{\left(\dfrac{W}{L}\right)_{PM2}}{\left(\dfrac{W}{L}\right)_{NM4}} = \frac{u_n}{u_p} \frac{2\left(V_{dd} - 1.5V_{tn}\right)V_{tn}}{\left(V_{dd} + V_{tp}\right)^2}$$

$$\text{Pull up ratio}(\gamma) = \frac{\left(\dfrac{W}{L}\right)_{PM2}}{\left(\dfrac{W}{L}\right)_{NM4}}$$

The desirable voltage required for proper writing operation at node Qb
after activation of WL is ΔV. As the voltage at node Q increases and then
drives the inverter PM$_2$ and NM$_2$, consequently the node Qb is pulled down
to 0V voltage through ON transistor NM$_2$. Hence, the cell state has been
changed. After performing the write operation, the WL becomes inactive. A
proper selection of pull-up ratio can guarantee a successful write operation.
A low value of pull-up ratio results in a low ΔV, and a low value of ΔV

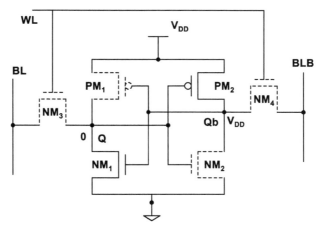

Figure 8.4 Hold operation of typical 6T SRAM cell.

provides high drive strength at the input of inverter NM_1-PM_1. To guarantee a low value of γ, the width of access transistor must be increased [7]. However, a wide access transistor will affect the read stability by lowering the cell ratio. Thus, there is a trade-off between proper performing read and write operation of the cell [4]. Various techniques used to overcome this conflict are discussed in the following section.

8.4 HOLD OR STAND-BY OPERATION

The operation of the SRAM cell in stand-by mode is depicted in Figure 8.4. In this mode, the 6T SRAM cell is in hold mode. Since no inputs are applied to the cell in this mode, the cell is isolated from all the input and output lines. As a result, the previously stored data remain in the cell. Owing to the various leakage current components in the SRAM cell, more static power is dissipated during the stand-by mode. The leakage current arises due to the OFF state of the transistors. Because of the leakage currents, the cell can lose its capability to hold data [3, 5, 8].

When designed at an advanced technology node [9, 10] for higher density and speed [11], the typical 6T SRAM cell faces various issues such as write stability, read ability, leakage current, static power dissipation, data retention, corners variation, and bit line leakage, among others. To date, tremendous efforts have been exerted to improve the performance of a SRAM cell. Some of these are discussed in this chapter to see the trends in SRAM cells. Figure 8.5(a) depicts the 7T SRAM cell, which was proposed in references [12–14]. As can be seen in Figure 8.5(a), this SRAM structure has a separate write and read port, thus keeping it free from read and write conflict. But this cell

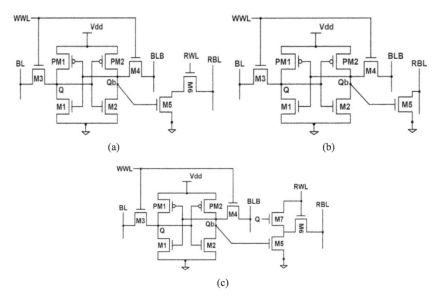

Figure 8.5 Different SRAM cell structures: (a) 7T SRAM cell; (b) 8T SRAM cell; (c) 9T SRAM cell.

suffers from another issues such as leakage, power dissipation, bit line leakage, etc.

8T SRAM architecture had been proposed in [15–18]. In this design, a bit interleaving enabled 8T SRAM cell is implemented. Figure 8.5(b) demonstrates such kind of arrangement. This SRAM cell has better performance in terms of power and area, i.e., both the power and occupied area are reduced in this cell. However, this design suffered from writing ability and data retention issues. A temperature-aware 8T SRAM cell had been proposed by N. Le Ba et al. [17] to verify the design at a high-temperature range. The write margin and read margin have not been affected much at high temperatures while more power is dissipated. A synchronous 2RW 8T dual-port SRAM cell has been proposed by Y. Yokoyama et al., for power reduction at the expense of increased read and write delay [18]. A 9T SRAM cell was designed by A. Teman et al. [19]. To improve the read stability, a separate read port has been included. A separate feedback path was used to improve the write stability. Nevertheless, the 9T SRAM cell suffers from half select and power dissipation problem. M. Tu et al. [20] designed a disturb-free 9T SRAM cell. To improve the read stability in this cell, a separate read port is used. A negative bit line technique has also been implemented to improve the write ability. This, however, resulted in the increase in overall chip area for SRAM. 9T SRAM cell by B. Wang et al. [21] is shown in Figure 8.5(c). The advantage of this 9T SRAM cell is that it can work with ultra-low voltage. It also uses a separate read port to enhance the read stability, but it suffers from write ability and delay issues.

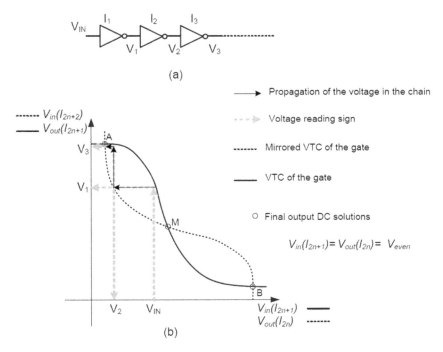

Figure 8.6 (a) Infinitely long chain of inverter. (b) VTC curve for various inverter.

8.5 SRAM PERFORMANCE PARAMETERS

8.5.1 Stability analysis of SRAM

The most important parameter in the SRAM cell is data stability. It analyses the robustness of SRAM cell as well as ensures the proper working in different operations [22]. The data stability concept serves as the foundation for realizing digital computing using semiconductor devices such as BJT and MOSFET. Essentially, it forms the connection between the physical voltage levels at the output and input of a CMOS inverter (or gate) to digital Boolean logic states [12]. As a result, a CMOS inverter can undertake a logical operation for any random static input, and the static output voltage for the inverter's series connected infinitely long chain converges to one of the three voltage levels associated with the inverter, as shown in Figure 8.6(a). The stability criteria have been verified by the voltage transfer characteristic (VTC) curve of the CMOS inverter. Figure 8.6(b) shows the VTC of cross-coupled CMOS inverter pairs. The second VTC curve is a mirror image of the first VTC curve. As given in Figure 8.6(b), if the VTC and its mirror (dotted line) of the inverter coincide at three points (A, B, and M), then the inverter chain output will be at VA, VB, and VM according to V_{IN}. Also, input and output are both at the same potential at point M [13]. The static noise margin (SNM) is a numerical measure of the ability of a system to generate logic states. The inverter connected in a long chain has some noise, which leads to the DC disturbance. The VTC curve will be affected by these DC noise sources. The noise sources can be added in the inverter either from the input side or through power supply of inverter [14].

There must be a noise margin for every sort of noise source. It has also been observed that the behavior of the noisy inverter chain can be replaced by a loop made up of the inverters in the chain.

8.6 POWER DISSIPATION IN SRAM CELL

The most crucial factor for SRAM in the nanoscale range is power dissipation [23]. Primarily there are two types of power dissipation in SRAM: dynamic power and static power [24]. The following section will give a brief idea about these power dissipations.

8.6.1 Dynamic power dissipation

The switching activities in the SRAM cell are the main cause of the dynamic power dissipation. Dynamic power can further be divided in two categories: switching power and short circuit power [25]. The concept of switching power can be understood by the CMOS inverter. The inverter is having some input and some load capacitance at output. During the transition of input signal, the load capacitor is charge and discharge. During the charging of the load capacitor, the energy is drawn from V_{DD}. Fifty percent of energy drawn is stored in a load capacitor while the remaining 50% is dissipated as heat in the pull-up network. During the discharging, the stored charge in the load capacitor is discharged into the ground. Hence in one complete cycle there is a loss of half the power, which is represented as switching or dynamic power dissipation [26]. Equation (8.3) gives the switching power:

$$P_{Switching} = \alpha C_L V^2_{DD} f_T \tag{8.3}$$

where C_L is the total load capacitance, including parasitic, α is activity factor, V_{DD} is supply voltage, and f_T is transition frequency.

8.6.2 Static power dissipation

During steady state the static power dissipates in an electronics circuit. This happens because a static current flows from the power source to the ground when there is no switching activity. Since in a CMOS circuit the pull-up and pull-down networks are not turned ON simultaneously during steady-state conditions, ideally the power dissipation in such a circuit will be zero.

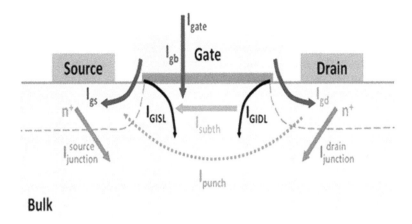

Figure 8.7 Various leakage currents in MOS [15].

In practical terms, however, this is not possible. However, in a CMOS circuit, various devices are turned OFF, hence there are a number of leakage currents exist, as shown in Figure 8.7 (i.e., sub-threshold leakage current, gate leakage current, gate-induced drain leakage current, punch-through current, junction leakage current, etc.). In nanometer technology, leakage current becomes a critical issue.

8.7 SRAM DESIGN CHALLENGES AT LOWER TECHNOLOGIES

The SRAM possesses many challenges when designed at a lower technology node. At such technology, supply voltage as well as threshold voltage of the device are scaled down. In other words, the devices operate under the sub-threshold region. The major design challenges are as follows:

- Reduction in current ratio (I_{ON}/I_{OFF})
- Decreased static noise margin
- Impact of PVT
- Leakage-dominated power
- Soft error

As compared to the super-threshold region, the SRAM faces critical challenges in the sub-threshold region. Higher variation across process corners and a significant reduction in I_{ON}/I_{OFF} ratio have resulted in a decrease in SRAM performance and stability [27]. The current ratio can be varied by 1000x for different process corners. The functionality of an SRAM circuit is heavily influenced by the relative strength of the devices used in the circuit.

Therefore, the read-and-write failure is considerably increased owing to the variation in I_{ON}. However, in the sub-threshold region leakage current (I_{OFF}) rises, affecting the power and reliability of SRAM. The reduction in the I_{ON} current affects the stability and reliability of SRAM [28]. SRAM stability is defined in terms of SNM. The SNM decreases with decreasing voltage, which indicates less stability and reliability. Under sub-threshold, the leakage current is more dominant which offers high power dissipation in the SRAM circuit. With excessive voltage scaling in the sub-threshold region, the charge at the storing node, which is responsible for holding the data state, is decreased, resulting in the data state being flipped due to a smaller particle strike. This increases the risk of device failure due to soft error, which is based on the location and usage. This soft error also affects the reliability of SRAM [29].

8.8 10T SRAM CELL USING TRI-STATE BUFFER

Figure 8.8 depicts the 10T SRAM cell made up of six NMOS and four PMOS transistors. Four transistors (PM1, PM2, NM1, and NM2) make the latch of the SRAM cell to store the data, whereas NM3 and NM4 act as access transistors. During the write operation, the access transistors are initiated by the write word line (WWL). The read port is made up of transistors PM3, PM4, NM5, and NM6 that are used to isolate the RBL from the storing nodes Qb and Q of the SRAM cell in order to provide a disturbance-free read operation. In read operations, transistors NM5 and PM4 serve as access transistors for SRAM cells, which are managed by read word line (RWL) and read word linebar (RWLB). During write operation the write bit line (WBL) and write bit linebar (WBLB) are the input lines

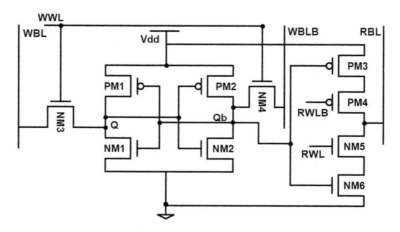

Figure 8.8 Schematic diagram of 10T cell design.

of a cell, whereas RBL is the input or output line during a read operation. This design's pull-up and cell ratio are retained at 0.6 and 1.5, respectively. In this design the read port is constructed by a tri-state buffer circuit, which consists of two PMOS and two NMOS transistors connected as a series, which provides high current and voltage amplification. The bit line leakage has been reduced in the read operation by using a tri-state buffer at the read port of the 10T cell design.

8.9 CONCLUSION

With the advent of the Internet of Things (IoT) and System on Chip (SOC), the need for SRAM is increasing at a very high rate. For processing the information, SRAM is an essential part of processor architecture and plays a crucial role in recent SOCs by occupying a major portion of the overall silicon area. Furthermore, large memory arrays made up of SRAMs consume a significant amount of overall SOC power. Thus, it is desirable to have SRAM operations at a lower supply voltage for longer battery life in portable devices. In this chapter challenges in the design of SRAM using nanoscale CMOS are discussed. Various typical SRAM cells have been discussed here for IoT applications.

REFERENCES

[1] M. E. Sinangil, Y. Lin, H. Liao, and J. Chang, "A 290-mV, 7-nm Ultra-Low-Voltage One-Port SRAM Compiler Design Using a 12T Write Contention and Read Upset Free Bit-Cell," in *IEEE Journal of Solid-State Circuits*, vol. 54, no. 4, pp. 1152–1160, 2019, doi: 10.1109/JSSC.2019.2895236

[2] S. Ohbayashi et al., "A 65-nm SoC Embedded 6T-SRAM Designed for Manufacturability With Read and Write Operation Stabilizing Circuits," in *IEEE Journal of Solid-State Circuits*, vol. 42, no. 4, pp. 820–829, 2007, doi:10.1109/JSSC.2007.891648

[3] G. Torrens et al., "A 65-nm Reliable 6T CMOS SRAM Cell with Minimum Size Transistors," *IEEE Transactions on Emerging Topics in Computing*, 2017.

[4] M. Khellah et al., "Process, Temperature, and Supply-Noise Tolerant 45\$~\$nm Dense Cache Arrays With Diffusion-Notch-Free (DNF) 6T SRAM Cells and Dynamic Multi-Vcc Circuits," in *IEEE Journal of Solid-State Circuits*, vol. 44, no. 4, pp. 1199–1208, 2009, doi:10.1109/JSSC.2009.2014015

[5] V. P. Hu, M. Fan, P. Su and C. Chuang, "Analysis of Ultra-Thin-Body SOI Subthreshold SRAM Considering Line-Edge Roughness, Work Function Variation, and Temperature Sensitivity," in *IEEE Journal on Emerging and Selected Topics in Circuits and Systems*, vol. 1, no. 3, pp. 335–342, 2011, doi:10.1109/JETCAS.2011.2163691

[6] V. P. Hu, "Reliability-Tolerant Design for Ultra-Thin-Body GeOI 6T SRAM Cell and Sense Amplifier," in *IEEE Journal of the Electron Devices Society*, vol. 5, no. 2, pp. 107–111, 2017, doi:10.1109/JEDS.2016.2644724

[7] J. Rabaey, A. Chandrakasan, and B. Nicolic, *Digital Integrated Circuits A Design Perspective*, 2nd ed. Prentice Hall, 2003.

[8] S. Gupta, K. Gupta and N. Pandey, "A 32-nm Sub-threshold 7T SRAM Bit Cell With Read Assist," in *IEEE Transactions on Very Large Scale Integration (VLSI) Systems*, vol. 25, no. 12, pp. 3473–3483, 2017.

[9] Y. Yang, H. Jeong, S. C. Song, J. Wang, G. Yeap, and S. O. Jung, "Single Bit-Line 7T SRAM Cell for Near-Threshold Voltage Operation With Enhanced Performance and Energy in 14 nm FinFET Technology," in *IEEE Transactions on Circuits and Systems I: Regular Papers*, vol. 63, no. 7, pp. 1023–1032, 2016.

[10] D. Sachan, H. Peta, K. S. Malik, and M. Goswami, "Low power multi threshold 7T SRAM cell," *2016 IEEE 59th International Midwest Symposium on Circuits and Systems (MWSCAS)*, Abu Dhabi, 2016, pp. 1–4.

[11] G. Samson, N. Ananthapadmanabhan, S. A. Badrudduza, and L. T. Clark, "Low-Power Dynamic Memory Word Line Decoding for Static Random Access Memories," in *IEEE Journal of Solid-State Circuits*, vol. 43, no. 11, pp. 2524–2532, 2008, doi:10.1109/JSSC.2008.2005813

[12] R. Saeidi, M. Sharifkhani and K. Hajsadeghi, "A Subthreshold Symmetric SRAM Cell With High Read Stability," in *IEEE Transactions on Circuits and Systems II: Express Briefs*, vol. 61, no. 1, pp. 26–30, 2014, doi:10.1109/TCSII.2013.2291064

[13] J. Lohstroh, E. Seevinck, and J. D. Groot, "Worst-case Static Noise Margin Criteria for Logic Circuits and Their Mathematical Equivalence," *IEEE Journal of Solid-State Circuits*, vol. SC-18, pp. 803–807, 1983.

[14] E. Seevinck, F. List, and J. Lohstroh, "Static-noise Margin Analysis of MOS SRAM cells," *IEEE Journal of Solid-State Circuits*, vol. SC-22, pp. 748–754, 1987.

[15] L. Wen, X. Cheng, K. Zhou, S. Tian, and X. Zeng, "Bit-Interleaving-Enabled 8T SRAM With Shared Data-Aware Write and Reference-Based Sense Amplifier," in *IEEE Transactions on Circuits and Systems II: Express Briefs*, vol. 63, no. 7, pp. 643–647, 2016.

[16] Ruchi, and S. Dasgupta, "Compact Analytical Model to Extract Write Static Noise Margin (WSNM) for SRAM Cell at 45-nm and 65-nm Nodes," in *IEEE Transactions on Semiconductor Manufacturing*, vol. 31, no. 1, pp. 136–143, 2018.

[17] N. Le Ba, and T. T. H. Kim, "Design of Temperature-Aware Low-Voltage 8T SRAM in SOI Technology for High-Temperature Operation (25 %C–300 %C)," in *IEEE Transactions on Very Large Scale Integration (VLSI) Systems*, vol. 25, no. 8, pp. 2383–2387, 2017.

[18] Y. Yokoyama, Y. Ishii, H. Okuda, and K. Nii, "A dynamic power reduction in synchronous 2RW 8T dual-port SRAM by adjusting wordline pulse timing with same/different row access mode," *2017 IEEE Asian Solid-State Circuits Conference (A-SSCC)*, Seoul, 2017, pp. 13–16.

[19] A. Teman, et al., "A 250 mV 8 kb 40 nm Ultra-Low Power 9T Supply Feedback SRAM (SF-SRAM)," *IEEE Journal of Solid-State Circuits*, vol. 46, no. 11, pp. 2713–2726, 2011.

[20] M. Tu, et al., "Single-Ended Subthreshold SRAM with Asymmetrical Write/Read-Assist," *IEEE Transactions on Circuits and Systems I: Regular Papers*, vol. 57, no. 12, pp. 3039–3047, 2010.

[21] B. Wang, et al., "Design of an Ultra-Low Voltage 9T SRAM with Equalized Bit-line Leakage and CAM-Assisted Energy Efficiency Improvement," *IEEE Transactions on Circuits and Systems I: Regular Papers*, vol. 62, no. 2, pp. 441–448, 2015.

[22] A. Makosiej, O. Thomas, A. Amara and A. Vladimirescu, "CMOS SRAM Scaling Limits Under Optimum Stability Constraints," *2013 IEEE International Symposium on Circuits and Systems (ISCAS)*, 2013, pp. 1460–1463, doi:10.1109/ISCAS.2013.6572132

[23] K. Bong, S. Choi, C. Kim, D. Han and H. J. Yoo, "A Low-Power Convolutional Neural Network Face Recognition Processor and a CIS Integrated With Always-on Face Detector," in *IEEE Journal of Solid-State Circuits*, vol. 53, no. 1, pp. 115–123, 2018.

[24] G. Samson, N. Ananthapadmanabhan, S. A. Badrudduza and L. T. Clark, "Low-Power Dynamic Memory Word Line Decoding for Static Random Access Memories," in *IEEE Journal of Solid-State Circuits*, vol. 43, no. 11, pp. 2524–2532, 2008, doi:10.1109/JSSC.2008.2005813

[25] G. Razavipour, A. Afzali-Kusha, and M. Pedram, "Design and Analysis of Two Low-Power SRAM Cell Structures," in *IEEE Transactions on Very Large Scale Integration (VLSI) Systems*, vol. 17, no. 10, pp. 1551–1555, 2009, doi:10.1109/TVLSI.2008.2004590

[26] P. Upadhyay, R. Kar, D. Mandal, and S. P. Ghoshal, "A Design of Low Swing and Multi threshold Voltage based Low Power 12T SRAM Cell," in *Computers & Electrical Engineering*, vol. 45, pp. 108–121, 2015.

[27] H. N. Patel, F. B. Yahya, and B. H. Calhoun. "Optimizing SRAM bitcell reliability and energy for IoT applications," in *2016 17th International Symposium on Quality Electronic Design (ISQED)*, pp. 12–17, March 2016.

[28] Nayak, D., Acharya, D. P., and Mahapatra, K. (2017). Current starving the SRAM Cell: a strategy to improve cell stability and power. *Circuits, Systems, and Signal Processing*, 36(8), 3047–3070.

[29] Malagón, D., Bota, S. A., Torrens, G., Gili, X., Praena, J., Fernández, B., ... Segura, J. (2017). Soft Error Rate Comparison of 6T and 8T SRAM ICs using Mono-energetic Proton and Neutron Irradiation Sources. *Microelectronics Reliability*, 78, 38–45.

Chapter 9

Variants-based gate modification (VGM) technique for reducing leakage power and short channel effect in DSM circuits

Uday Panwar
SIRT, Bhopal, India

Ajay Kumar Dadoria
Amity University, Gwalior, India

9.1 INTRODUCTION

As demand for portable devices continues to increase, manufacturers have been following the Moore's law of scaling in last decade for increasing the device complexity and improving functionality with upcoming technologies. Rapid-integration, high-speed, and low-cost ultra large-scale integration (ULSI) circuit has recently been added for mitigation of power consumption by various circuit designs used today. For the complementary metal oxide semiconductor (CMOS) technology, power dissipation is the major concern in complex circuit design [1, 2].

Short channel devices are those in which the gate lengths are scaled to such an extent that it becomes comparable to the source and drain depletion region thicknesses. The prominent SCEs in the devices include drain-induced barrier lowering (DIBL) [3], mobility degradation, carrier velocity saturation, and punch-through. These SCEs lead to an increase in sub-threshold leakage, reduction in drive current, sub-threshold slope degradation, and decrease in MOSFT reliability [4]. Thus, SCEs put a limit on scaling of gate length in MOSFETs, which forces the VLSI scientists to think of an alternate non-planer technology called Fin-FETs.

Fin-FET was invented by Chenming Hu in 1998 at the University of California–Berkeley. Intel made the first tri-gate Ivy Bridge processor at 22 nm in 2012 [5]. Due to its reduced fringing and parasitic capacitances, Fin-FET offers reduced area, power consumption, and SCEs along with an improved speed of operation [5, 6]. Fin-FET can be either silicon on insulator (SOI), or bulk substrate depending on the application. The SOI substrate is used more frequently, though. Figure 9.1 shows the three-dimensional (3D) structure of Fin-FET in which a silicon fin stands vertically on the SOI/ bulk substrate.

DOI: 10.1201/9781032670270-9

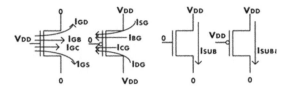

Figure 9.1 Representation of leakage current characteristic.

The working of Fin-FET is guided by the same set of conditions as that of planer MOSFET. Gate, drain, and flat band in DG Fin-FET are defined as V_{gs}, V_{ds}, and V_{fb}, respectively. In n-channel Fin-FET, if $V_{gs} < V_{fb}$, then it does not conduct, and if $V_{gs} > V_{fb}$, channel inversion occurs. When $V_{gs} > V_{fb}$ and $V_{ds} < V_{ds-sat}$, the device is in linear mode, and when $V_{gs} > V_{fb}$ and $V_{ds} > V_{ds-sat}$, the device is in saturation. In DG Fin-FET, the current from drain to source in a linear region (I_{ds-Lin}), when $V_{ds} < V_{ds-sat}$ is given by Equation (9.2),

$$I_{ds-Lin} = 2\mu \left(\frac{w}{L}\right) C_{ox} \left[\left(V_{gs} - V_{th} - \frac{V_{ds}}{2}\right) V_{ds}\right] \qquad (9.1)$$

where C_{ox} is the oxide capacitance and V_{th} in the above equation by the Equation (9.2),

$$V_{th} = V_0 + 2V_t \ln\left[\frac{V_{gs} - V_0}{4rV_t}\right] \qquad (9.2)$$

V_t in Equation (9.2) is the thermal voltage given by Equation (9.3),

$$V_t = \frac{KT}{q} \qquad (9.3)$$

where K is Boltzmann constant, T is the temperature in Kelvin, and q is the electronic charge V_0 in Equation (9.2) is a parameter in FinFET defined by Equation (9.4),

$$V_0 = V_{fb} + 2V_t \ln\left[\frac{2}{F_t}\right] \sqrt{\frac{2\epsilon_{si}V_t}{qn_i}} \qquad (9.4)$$

and the parameter r in Equation (9.3) is equal to the ratio given in Equation (9.5),

$$r = \frac{\epsilon_{si}t_{ox}}{\epsilon_{ox}F_t} \qquad (9.5)$$

The expression for I_{ds-Lin} is the same as that in planer MOSFET, and the multiplication of the factor of 2 to account for the current flow from two sides of the fin.

Similarly, in DG Fin-FET, when $V_{ds} > V_{ds-sat}$ in the saturation region, (I_{ds-sat}) is given by Equation (9.6),

$$I_{ds-sat} = \mu\left(\frac{w}{L}\right)C_{ox}\left[\left(V_{gs} - V_{th}\right)^2 - 8rV_t^2 e^{\left(\frac{V_{gs} - V_0 - V_{ds}}{V_t}\right)}\right] \tag{9.6}$$

As Fin-FET also leaks in the subthreshold region although very less as compared to planner devices, the expression of the I_{sub} is given by Equation (9.7) [7].

$$I_{sub} = \mu\left(\frac{w}{L}\right)KTF_t n_i e \frac{\left(V_{gs} - V_{fb}\right)}{V_t}\left[1 - e^{\frac{V_{ds}}{V_t}}\right] \tag{9.7}$$

The value of I_{sub} is kept at minimum by adjusting the values of F_t [8], as the body in Fin-FET is ultrathin and is completely depleted of charge carriers. This makes the gate control from more than one side very significant with reduced value of I_{sub}.

The chapter is organized as follows: Section 9.2 gives an overview of certain previous works of the leakage control techniques. Section 9.3 explains the proposed technique. Section 9.4 elucidates results and discussion of the different circuits. Section 9.5 concludes the paper.

9.2 MOTIVATION

Leakage in complementary metal oxide semiconductor (CMOS) is a significant issue in deep sub-micron (DSM) technology, where gate oxide leakage current (IGATE) and ISUB play a crucial role. Leakage current has various components, and Figure 9.1 illustrates the properties of IGATE. For both N-type and P-type metal oxide semiconductor (NMOS and PMOS) devices, the gate-to-substrate leakage current caused by electron tunneling from the valence band is referred to as IGB. IGS and IGD refer to gate-to-source and gate-to-drain overlap leakages, respectively. During an inversion region, gate-to-channel tunneling current is denoted by IGC. These currents reach their maximum values due to specific circumstances [9].

$$I_{SUB}\left(MAX\right): V_{DS} = V_{DD}\left(1\right)$$

$$I_{GATE}\left(MAX\right): V_{GS} = V_{GD} = V_{GB} = V_{DD}$$

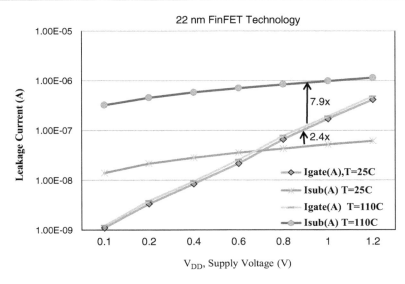

Figure 9.2 Variation in I_{SUB} and I_{GATE} for different supply voltages at T = 25°C and 110°C temperatures.

where V_{GS}, V_{GD}, and V_{GB} are the gate-to-source, drain, and substrate voltages, respectively; V_{DS} is the drain-to-source voltage; and V_{DD} is the supply voltage. The bias condition for the NMOS and PMOS transistors that results in the highest I_{GATE} and I_{SUB} currents is depicted in Figure 9.2 [10]. Here, the first two and the last two transistors, respectively, reflect the maximum I_{SUB} and I_{GATE} conditions.

Figure 9.2 illustrates how the I_{SUB} and I_{GATE} change with temperature and V_{DD} in a 22-nm Fin-FET technology. While the I_{GATE} exhibits a less significant dependency on temperature, I_{SUB} grows exponentially as the temperature rises. At the standard supply voltage ($V_{DD} = 1V$), the NMOS transistor's I_{SUB} is 7.9 times bigger at 110°C than the I_{GATE}. In contrast, I_{GATE} is 2.4 times larger than the I_{SUB} at ambient temperature [11, 12].

So the I_{GATE} majorly depends on two parameters of the device: SCEs and DIBL effects. This became the primary leakage source in DSM circuits.

This work aims to address two issues related to deep submicron (DSM) circuits: high power consumption and leakage. To mitigate these problems, Fin-FET devices are used in combination with the input vector control (IVC) technique. IVC forces the circuit's primary input to its minimum leakage vector (MLV) during idle state. This method considers leakage at each input vector, and selects the vector with the least amount of leakage as the MLV. Table 9.1 shows the leakage behavior of NAND logic circuits with two and three inputs, respectively, simulated using the 45-nm Berkeley Predictive Technology Model (BPTM) [13]. The input vector has a direct impact on the leakage of the logic circuit, with '00' being the MLV and '11' being the worst leakage state (WLS) of the two-input NAND logic. MLV saves 97.68% of

Table 9.1 Leakage behavior for NAND logic

S.No.	Input vector	Leakage current (pA)
(a) 2 input NAND		
1	0	1.29
2	1	12.34
3	10	4.34
4	11	18.35
(b) 3 Input NAND		
1	0	1.21
2	1	8.34
3	10	1.38
4	11	21.35
5	100	1.28
6	101	16.36
7	110	12.39
8	111	32.37

the leakage power compared to the two-input NAND logic WLS. Therefore, MLV is chosen as the primary input in the circuit's idle state to minimize leakage power. However, the IVC approach may not be effective for reducing leakage when the logic depth is high. To address this limitation, this chapter proposes the variant-based gate modification (VGM) mechanism, which employs a three-input NAND gate instead of a two-input NAND gate to achieve greater leakage reduction.

9.3 LEAKAGE REDUCTION BY VARIANT-BASED GATE MODIFICATION

The conventional IVC technique doesn't effectively reduce leakage or control intermediate nodes in a circuit. In this work, we propose an algorithm that modifies logic gate terms using the VGM technique, which reduces leakage at each level and helps reduce SCEs and DIBL [14] in Fin-FET technology. Section 9.3.1 describes the conventional gate replacement (GR) technique, while Section 9.3.2 presents our VGM algorithm in pseudo code. We test our algorithm on a benchmark circuit, specifically C17 (ISCAS'85) [8, 11, 12, 15], in Section 9.3.3.

9.3.1 Typical gate replacement technique

The conventional GR technique uses an additional input dynamically when a logic gate enters its WLS. This technique reduces the circuit's leakage power when it is in stand-by mode. To represent this approach in the

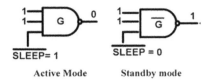

Figure 9.3 Traditional gate replacement technique employed in worst leakage state condition during stand-by mode.

circuit, we can consider a gate N(i) in the idle state, which can be replaced by another gate N (i, SLEEP) for a specific input vector i. The sleep signal helps limit leakage in the circuit when it's in standby mode (see Figure 9.3). This approach doesn't affect the logic gate's operation, which remains the same as in its active mode.

In the case of standby mode the SLEEP = 1

N (i, SLEEP) = N (i, 0) = N (i) with decreased leakage current.

In the case of active mode SLEEP = 0

N (i, SLEEP) = N (i, 1) = N (i) with the proper functionality of the circuit.

9.3.2 Proposed algorithm for VGM technique

Some NAND gate variations, which are slightly modified of the original gate already incorporated into the standard cell library, have been designed with the objective of minimizing the leakage of the IC. These variations have similar characteristics like traditional two-input NAND gate. With a WLS condition, this variant is substituted by the original gate to reduce leakage current while maintaining the same fan-out. Because it is costly and time consuming, this procedure of developing variations is only done once before designing the circuit. Figure 9.5 displays the Fin-FET NAND variations discussed above.

In the context of WLS for a two-input NAND gate, the application of variants V1-V4 instead of the WLS gate can reduce the number of OFF state transistors required, from '2' to '1', '1' to '0', and '0', respectively (for V1-V4). During idle mode, the OFF state transistors must meet specific conditions, such as being in cut-off mode and having a maximum VDS. Table 9.1(a) shows that the WLS for a two-input NAND gate is "11", in which case both PMOS transistors are in the OFF condition, satisfying the maximum criteria of ISUB. To reduce the number of OFF-condition PMOS transistors in this scenario, we can use the proposed variants. Figure 9.4 shows variations V1, V2, and V4, which can be utilized as replacements once a device enters its idle state. The benefit of these specific variants is that the output characteristics of the WLS transistor do not change, and therefore a succeeding transistor associated with the output of the variant does not require any replacement and remains unchanged.

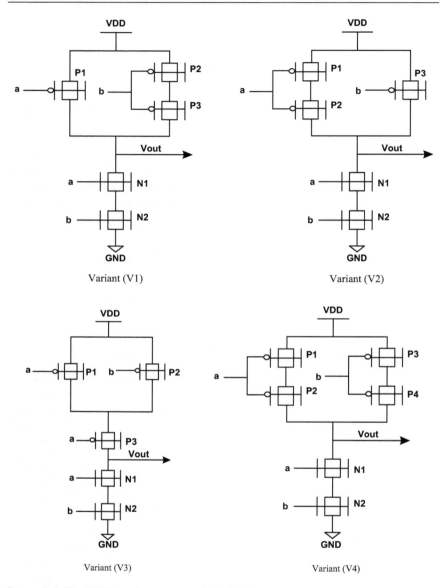

Figure 9.4 Fin-FET based variants as V1, V2, V3, and V4 used in VGM technique.

In contrast, the conventional GR technique, as described in Section 9.3.1, replaces a two-input NAND gate with a three-input NAND gate, which is contradictory. Variant V4 is useful when greater leakage reduction is a priority. It reduces the number of OFF state transistors forming '2' to '0' through a single PMOS transistor connecting a pull-down network (PDN) and an output. However, it slightly increases the delay of the gate.

Proposed algorithm for VGM

Begin:

---

1. INPUT (levelized logic gate list, SCD) //Standard Cell Based Library
2. OUTPUT (optimized circuit in terms of leakage power dissipation with unchanged characteristics in the switching state)
3. Primary inputs of logic circuits set to their minimum leakage vector state (as per given logic gate at 0th level.

For each logic level from 1 onward

4. Step 1- For (Level 1 = 1; Level 1 <= n (total number of level of the circuit); L1++);
5. Step 2: For (G = 1; G<= k (number of gates at Level1);G++);
6. If (Logicstate = WLS) // let say for NAND = 11 as WLS;
7. Replace it with its VGM // NAND VARIANT V1-V4;
8. Go to step (2);
9. Else Logic is at MLV then without apply VGM just go to step (2);
10. End if;
11. End for;
12. End for

9.3.3 Proposed algorithm applied on C17 (ISCAS'85) Benchmark circuit

Here proposed VGM technique applied and tested on C17 benchmark circuit as represented in Figure 9.5(a) Conventional NAND-based C17 and (b) modified C17 using VGM-based variant V3.

After applying the VGM technique on given C17 circuit, it reduces total leakage current of 76% of replace gates as displayed by V3 as compared with conventional C17 circuit.

9.4 RESULT AND DISCUSSION

The VGM technique, which is a variation-based gate modification method, was tested on various benchmark circuits such as C17-ISCAS'85, B01,

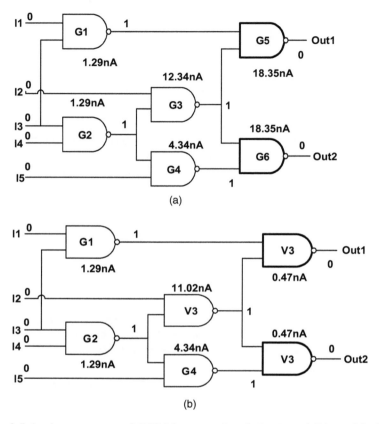

Figure 9.5 Leakage current of C17 (a) conventional circuit and (b) modified C17 circuit.

B02, and B06-ITC'99 using CMOS gates. The study included analyzing the CMOS NAND gate and its four different versions (V1 to V4), and the results showed the amount of leakage current as well as the total power (PT), which is a combination of static power dissipation (PST) and dynamic power dissipation (PDY). Tables 9.5 and 9.6 present the results for PST and PDY, respectively.

Table 9.2 (a) NAND gate with variation (V1) and (V2)

S. No.	Leakage current (pA)	Total power (P_T (nW))	Delay (fS)
1	1.29	15.25	67.98
2	12.34	16.23	
3	4.34	15.76	
4	9.24	16.98	

Table 9.3 (b) NAND gate with variation (V3)

S. No.	Leakage current (pA)	Total power (P_T (nW))	Delay (fS)
1	2.56	10.13	86.67
2	3.43	10.48	
3	7.67	10.96	
4	0.965	11.24	

Table 9.4 (c) NAND gate with variation (V4)

S. No.	Leakage current (pA)	Total power (P_T (nW))	Delay (fS)
1	1.29	12.35	98.67
2	11.02	13.33	
3	7.67	12.86	
4	0.476	13.99	

9.4.1 Discussion on leakage current

Table 9.5 illustrates the results of direct current analysis on various benchmark circuits and logic gates concerning leakage current using the 32-nm BPTM [16] file, with a voltage of 1V and a temperature of 25°C. The PMOS and NMOS transistor width-to-length (W/L) ratios were 3 and 1.5, respectively. The proposed algorithm was implemented on ten circuits, and Table 9.5 presents the corresponding leakage current (in pA). For the NAND gate, the variant V3 had a leakage current of 1.29 pA at '00' input. According to the proposed algorithm, the input replacement with variant V3 was not

Table 9.5 Leakage current (pa) with different logic gates and benchmark circuits at 32-nm finfet technology

			Circuit if replace by variation according to proposed algorithm		
Leakage current (pA)					
S. No.	Logic design	Conventional design	V1 & V2	V3	V4
1	NAND	1.29	1.29	1.29	1.29
2	OR	55.45	28.95	3.05	2.80
3	AND	55.25	27.72	1.825	1.573
4	Ex-OR	80.65	54.15	37.05	28.08
5	Ex-NOR	81.87	55.37	28.25	29.22
6	1 Bit FA	214.3	134.8	29.475	56.32
7	C17 (ISCAS 85)	134.8	81.87	30.05	29.55
8	B01 (ITC 99)	1916.4	1042.5	420.67	414.7
9	B02 (ITC 99)	1062.5	584.28	195.6	191.6
10	B06 (ITC 99)	2203.8	1212.4	383.3	375.32

Table 9.6 Total power (P_T) (combination of static power dissipation (PST) and dynamic power dissipation (PDY)) with different logic gates and benchmark circuits at 32-nm FinFET technology

		Total power P_T (nW)			
S. no.	Logic design	Conventional design	VI & V2	V3	V4
I	NAND	2.66	3.57	3.36	3.45
2	OR	5.04	4.95	4.56	4.80
3	AND	4.53	4.47	4.26	4.44
4	Ex-OR	6.42	6.15	4.89	6.36
5	Ex-NOR	7.44	7.26	5.01	7.41
6	I Bit FA	8.31	7.41	6.72	7.32
7	C17 (ISCAS 85)	6.57	6.27	5.64	5.73
8	B01 (ITC 99)	763.5	716.88	780.84	716.3
9	B02 (ITC 99)	1277.1	822.78	531.42	682.32
10	B06 (ITC 99)	1433.9	1336.5	857.37	849.57

performed for '00,' and the leakage remained at 1.29 pA. The minimum leakage current was observed when the variant V4 was used as VGM for PT analysis.

The leakage current and static power dissipation order by variants is as follows:

V4 < V3 < V1&V2 < Conventional circuit

9.4.2 Total power dissipation analysis

P_T is the total power dissipation of a given circuit. The sum of the power in both analyses (i.e., direct current and transient) gives the P_T of the circuit.

To analyze PST, direct current analysis was conducted using a 32-nm BPTM [16] file at 1.1V and 25°C. The PMOS and NMOS transistors had width-to-length (W/L) ratios of 3 and 1.5, respectively. PDY was analyzed using transient analysis to measure switching or dynamic power dissipation for each variation. The input pulse signal conditions were as follows: pulse widths, pulse period 4 ns, rise and fall time 10 ps, and simulation time 20 ns. Table 9.6 shows that variation V3 had the lowest PT (in uW).

The total power dissipation order by variants is as follows:

V3 < V4 < V1&V2 < Conventional circuit

9.5 CONCLUSION

The present study investigated leakage current and total power analysis in the next-generation DSM technology, focusing on the dominant sources of leakage power dissipation, namely ISUB and IGATE. These sources increase the leakage in DSM VLSI designs when the circuits are in idle mode. To address this issue, a novel technique called variation-based gate modification is proposed. The technique uses four variants to effectively control the ISUB current of a device and reduce leakage current in idle mode. Simulation results showed that the proposed algorithm outperforms the conventional GR technique in terms of leakage reduction while maintaining the fan-out logic state of WLS gates. The proposed technique was tested on different logic circuits and benchmarks, including C17 (ISCAS'85) and B01, B02, and B06 (ITC'99) circuits, totaling ten circuits. Using the V1–V4 variants of the proposed technique, the maximum average leakage power was reduced to 37.3% with variant V3, and the maximum average leakage current was reduced to 80.53% with variants V4 and V3, compared to the conventional circuit values. Variants V1, V2, and V4 did not increase the area of the device due to a transistor-stacking effect. The proposed technique has applications in high-performance, low-power devices such as microprocessors, memory units, and portable devices where leakage is a major concern.

REFERENCES

[1] S. Borkar, "Obeying Moore's law beyond 0.18 micron", in *Proceedings of the IEEE International ASIC/SOC Conference*, pp. 26–31, September 2000.

[2] International Technology Roadmap for Semiconductors (ITRS) 2001, 2002, Courtesy.

[3] Moore's Law meets static power, Computer, December 2003, IEEE Computer Society.

[4] S.M. Kang, Y. Lablebici, *CMOS Digital Integrated Circuits: Analysis and Design*, 3rd Edition, Tata Mc-Graw Hill, 2003.

[5] Ajay Kumar Dadoria, Kavita Khare, Tarunkumar Gupta, Uday Panwar, "Efficient Flipped Drain Gating Integrated with Power Gating Technique for FinFET Based Logic Circuit", *WILLEY-IJNM*, 31(5), 1–14, 2018. https://doi.org/10.1002/jnm.2344

[6] Ajay Kumar Dadoria, Kavita Khare, Uday Panwar, Anita Jain. "Performance Evaluation of Domino Logic Circuits for wide fan-in gates with FinFET", *Microsystem Technologies*, 24(8), 3341–3348, 2018. https://doi.org/10.1007/s00542-017-3691-3

[7] Uday Panwar, Kavita Khare, "Gate Replacement with PMOS stacking for Leakage reduction in VLSI Circuits", *Willey-IJNM*, 29(4), 565–576, 2015. https://doi.org/10.1002/jnm.2112

[8] B. S. Deepak Subramanyan, A. Nunez, "Analysis of Subthreshold Leakage Reduction in CMOS Digital Circuits", in *Proceedings of the 13th NASA Symposium*, Post Falls Idaho, pp. 1400–1404, June 5–6, 2007.

[9] Uday Panwar, Parul Sharma, "Design and Implementation of Ternary Adder for High-Performance Arithmetic Applications by Using CNTFET Material", in *Materials Today*, Vol. 63, pp. 773–777, 2022. https://doi.org/10.1016/j.matpr.2022.05.316

[10] Uday Panwar, Anat Shrivastava, "A Novel Technique to improve Performance Evaluation of Domino Logic Circuits in CMOS and FinFET Technology", *IEEE 2nd International Conference on Data, Engineering and Applications 2020 (IDEA-2k20)*, February 28–29, 2020.

[11] Rajeevan Chandel, S. Sarkar, R. P. Agarwal, "Performance Controlling Parameters of Voltage Scaled Repeaters for Long Interconnections", in *IETE Journal of Research*, Vol. 51, No. 2, pp. 107–113, 2015.

[12] B. K. Kaushik, "Transient Analysis of Hybrid Cu-CNT On-Chip Interconnects Using MRA Technique", in *IEEE Open Journal of Nanotechnology*, Vol. 3, pp. 24–35, 2022, 10.1109/OJNANO.2021.3138344

[13] Namrata Sharma, Uday Panwar, "A Novel Approach for Analysis CNTFET Based Domino Circuit in Nano Scale Design", in *International Journal of Advanced Science & Technology*, Vol. 29, No. 7, pp. 5898–5908, 2020.

[14] Anjali Patware, Uday Panwar, "A Novel Design CNTFET based Adiabatic Logic Circuit for Low Power Application" in *International Journal of Advanced Science & Technology*, Vol. 29, No. 7, pp. 5999–6011, 2020.

[15] A. P. Chandrakasan, R. W. Brodersen, *Low Power CMOS Digital Design*, Kluwer Academic: Norwell, MA, 1995.

[16] http://ptm.asu.edu/

Chapter 10

A novel approach for high speed and low power by using nano-VLSI interconnects

Narendra Kumar Garg, Vivek Singh Kushwah
and Ajay Kumar Dadoria
Amity University, Gwalior, India

10.1 INTRODUCTION

Interconnects are tiny wires made of conductive material that connect two or more nodes electrically. Gate delays are less noticeable at lower technology nodes as compared to connection delays. [1] At higher frequencies, interconnects don't just function as a simple resistor; they also have parasitic capacitance and inductance attached to them. These material-dependent impedance factors affect the speed, noise, delay, and power dissipation of circuits. Previously, the conductivity and SiO_2 adhesion of a material determined if it was appropriate as an interconnect. Due to electromigration caused by a rise in current density, traditionally utilized aluminum interconnects have been replaced by copper interconnects [2–6]. Copper continues to be the universally favored connecting material for industry. As a result of ongoing technological development and the scaling of semiconductor devices, copper began to experience an increase in resistivity due to surface roughness and gain boundary scattering. This calls for the use of cutting-edge connection materials, such as carbon nanomaterial [6–8]. By segmenting the lengthy link into smaller subsections and introducing smart buffers in between, propagation latency can be further reduced in addition to the use of appropriate interconnect materials. Buffer insertion technique is used to linearize the Elmore delay of unbuffered interconnect because it increases quadratically with an increase in wire length. In addition to speeding up the logic, repeaters and buffers can restore output waveforms that have been distorted by parasitic interconnect. However, the act of switching repeaters adds to overall latency and results in some power loss. Even when they are not switching, buffers can still use energy. Therefore, it becomes essential to provide intelligent repeaters or buffers that help increase interconnect speed without sacrificing static or dynamic power savings [9]. The main goal of this chapter is to gather data on the delay and power dissipation of compound repeaters in order to formulate the optimization of power delay product (PDP) and energy delay product (EDP) that are as efficient as possible based on CNT. Additionally, the propounded repeaters are contrasted with CNT and MOS technologies.

DOI: 10.1201/9781032670270-10

The remainder of the text is structured as follows. The modeling of interconnects and buffers based on CNT is described in Section 10.2. The approach taken into account when designing the driver interconnect load (DIL) test bench architecture is described in Section 10.3. Along with explaining how compound buffers function as repeaters, it also describes how ordinary buffers do the same. The findings of the simulation results are discussed in Section 10.4. Section 10.5 provides conclusions.

10.2 BACKGROUND AND PREVIOUS WORK

A carbon nanotube (CNT) is composed of carbon atoms that are arranged in sheets resembling benzene, which are then rolled to form single- or multi-walled tubes. CNTs have high conductivity compared to MOS technology. The direction in which the graphene sheet is rolled is represented by the chirality index (m, n) as shown in Figure 10.1. In CNTFET technology, when the chirality index is equal (m = n) or when the difference (m – n) is equal to 3 k where k is an integer, the conductivity of the device is maintained without affecting its properties. The conductive properties of the device increase due to the arrangement of carbon atoms and the formation of the channel of the transistor, either SWCNT or MWCNT channel. The diameter of the tube can be calculated using Equation (10.1) [10].

$$D_{CNT} = a\sqrt{\frac{m^2 + n^2 + mn}{\pi}} \tag{10.1}$$

In the above equation m and n represent the chirality vector and a is the lattice constant (2.49 Å). Figure 10.2 represents the top view of the sheet.

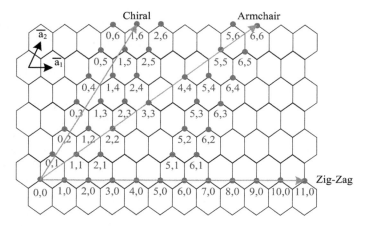

Figure 10.1 Two-dimensional structure of CNTFET sheet with arrangement of carbon atom.

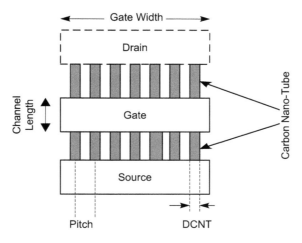

Figure 10.2 Top view of CNTFET.

The width of the CNTFET gate (W_{gate}) is calculated from Equation (10.2),

$$W_{gate} \approx \max\left(W_{min}, N \times Pitch\right) \qquad (10.2)$$

where Pitch is the distance between the tubes, W_{min} is minimum gate width, and N is the number of tubes.

The operation of a CNTFET is similar to MOS technology, but the key difference lies in the conductivity through the channel. CNTFETs have significantly higher conductivity than MOS technology. The channel is formed by enclosing intrinsic CNTs with a planar metal gate and a high-k dielectric such as zirconium oxide (ZrO_2) or hafnium oxide (HfO_2) as the gate oxide. The substrate is then fully covered by an insulating SiO_2 layer. In single-wall CNTFETs, multiple nanotubes are aligned in parallel according to the width of the gate, with the distance between the two tubes measured from their centers, called the pitch. In simulations, the source and drain are doped CNTs with a 0.8% doping level, and HfO_2 is used as the gate dielectric with a thickness of 3 nm, while the bulk dielectric SiO_2 has a thickness of 10 μm [11]. The threshold voltage is measured using the energy gap and charge across the channel,

$$V_{TH} = \frac{E_g}{2q} = \frac{1}{\sqrt[3]{3}} \frac{aV_\pi}{qD_{CNT}} \qquad (10.3)$$

$$E_g = \frac{0.872}{D_{CNT}} \text{ and } V_{TH} = \frac{0.436}{D_{CNT}} \qquad (10.4)$$

where D_{CNT} is in nm. For improvement of parameters I_{ON} should be increased so the conductivity improves.

10.2.1 Interconnect modelling

Both the circuit design and verification processes depend heavily on interconnect modeling. These processes can be improved by using a precise and effective connection model in conjunction with efficient CNT field effect transistor (CNTFET) buffers. Equivalent models of copper interconnect and interconnects based on carbon nanomaterials are given in the following subsections.

10.2.1.1 Modeling parameters of copper interconnects

Calculations of the parasitic impedance of global layer interconnect lines with coupling above the ground are done using the predictive technology model (PTM) model. The commonly used copper connector model with dimensions of thickness (t), length (l), width (w), and height (h) above ground plane is shown in Figure 10.1. The formula for rectangular copper interconnect's equivalent resistance is based on the following equation:

$$R = \rho l / wt$$

The measurement of resistance of the circuit is represented by R, and resistivity is represented by ρ. Different metals have different resistivity; here we are using Cu = 2.2 Ω cm.

For measurement of interconnect parameters like inductance, L is given as

$$L = \mu_0 \frac{1}{2\pi} \left[\ln\left(\frac{2l}{w+t}\right) + 0.5 + \left(\frac{0.22(w+t)}{l}\right) \right]$$

For measurement of the mutual inductance of M,

$$M = \mu_0 \frac{1}{2\pi} \left[\ln\left(\frac{2l}{d}\right) - 1 + \left(\frac{d}{l}\right) \right]$$

Capacitances encountered in a CNTFET are illustrated in Figure 10.3. These include C_{ox}, which is the intrinsic gate oxide capacitance, C_{do}, which is the direct capacitance in the gate-to-source overlapped region, and C_{if}, which is the inner fringing capacitance [12]. Capacitances that emerge from the sides of a poly-Si gate, pass through the sidewall, and terminate at the source/drain (S/D) regions are known as sidewall-fringing capacitances, or C_{of}. Additionally, C_{pp} is the parallel plate capacitance connected to the electric field lines that emerge from the polygate's sidewalls, pass through

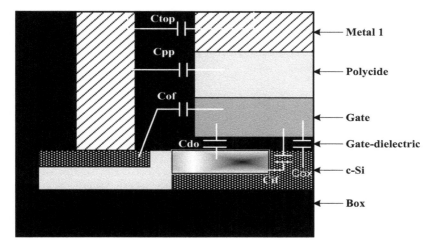

Figure 10.3 Various types of interconnects and capacitance formation in RLC circuits.

the sidewall spacer, and terminate at S/D electrodes (contact plugs) [13]. C_{top} is the top-fringing capacitance connected to field lines that emerge from the poly-top gate's surface, pass through the first layer of passivation/planarization dielectrics, and terminate at S/D electrodes (contact plugs) (Table 10.1).

The characteristics of four types of transistors, namely 32-nm N-Type MOS, SG mode FinFET, LP mode FinFET, and CNTFET, were analyzed. Simulation results indicate that CNTFET exhibits higher I_{ON} state current

Table 10.1 CNTEFT model parameters used in simulation

Parameters	Values
Physical Channel Length (L_{ch})	32 nm
Tube Diameter	$d = \dfrac{|C_h|}{\pi}$
Dielectric material of Gate thickness (t_{ox})	5 nm
Coupling Capacitor (C_{sub})	40 pF/m
CNT Work Function	4.6 eV
Pitch PCNTFET and NCNTFET	5 and 10
Lattice Constant	1.43 A°
V_{DD}	0.8 V
Chiral angle	$Tan\theta = \sqrt{(3m/((2n+m)))}$
Chiral vector	$C_h = na1 + ma2 = (n, m)$
The mean free path in p+/n+ doped CNT	15 nm
V_{Th} of the P-CFET and N-CFET	

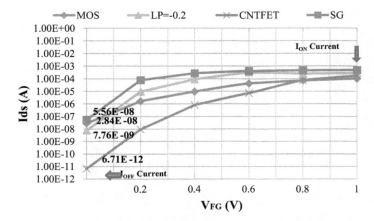

Figure 10.4 I-V characteristics of 32-nm MOS, FinFET (SG, LP mode), and CNTFET.

compared to MOSFET, resulting in stronger driving capability. Furthermore, CNTFET has lower I_{OFF} than both MOS and FinFET, leading to better suppression of leakage current. Consequently, CNTFET achieves faster switching speed, making it suitable for high-frequency applications, as illustrated in Figure 10.4.

Equivalent Diagram of RLC Circuits

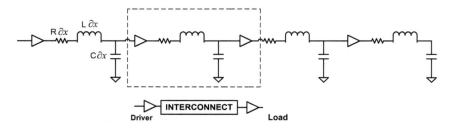

Figure 10.5 Equivalent circuit diagram of RLC interconnects.

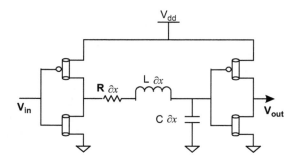

Figure 10.6 Convention RLC interconnects in inverters.

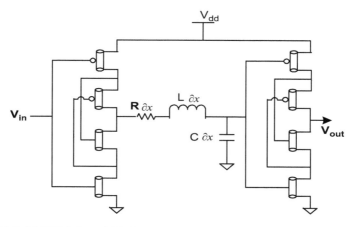

Figure 10.7 LECTOR-based RLC interconnects in inverters.

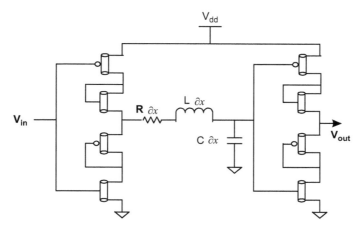

Figure 10.8 GALEOR-based RLC interconnects in inverters.

10.3 PROPOSED WORK

Similar to the NOT GATE logic, the functioning of the NMOS blocks is coupled in parallel with the PMOS transistors of the latch to form the transmission gates. However, there is a prime difference in the pull-down network, which includes an NMOS diode, the ground, and a transistor. The NMOS tree is utilized to manage the discharging path and reduce the rate of discharge of the internal nodes of the logic circuit. This method also incorporates the benefit of the suggested RLC interconnect strategy, which lowers the gate-to-source voltage at the output transistors, thus reducing leakage current and gate current.

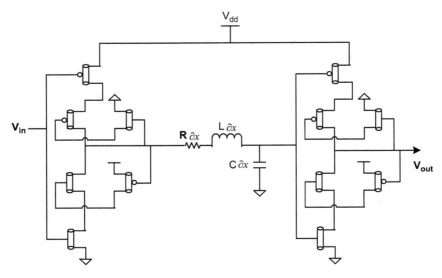

Figure 10.9 Proposed circuit.

To further improve the repeater's performance by reducing leakage current and addressing the issue of minimizing the delay, a self-controlled toggle switch block is inserted between the PUN and PDN. This block serves as a further reduction in leakage current and helps improve the repeater's performance.

When $V_{DS} > V_{GS} - V_T$ where V_T is threshold voltage, the circuit is in saturation region, which turns ON the transistor; this acts like a diode when $V_{GS} > = V_T$, and calculation of I_{DS} is given by

$$I_{DS} = K\left(V_{GS} - V_T\right)^2 = K\left(V_{DS} - V_T\right)^2 \qquad (10.5)$$

From the above equation drain current totally depends on the gate-to-source voltage and threshold voltage of the transistor.

In the proposed approach, the source terminal of MN3 is connected to positive DC voltage V_{dc}, which is connected to Gnd. Thus we see that the source voltage $VS = V_{dc}$. And so, $V_{DS} = VD - V_{dc}$. The equation can be represented as

$$I_{DS} = K\left(V_{DS} - V_T\right)^2 = K\left(\left(V_D - V_{dc}\right) - V_T\right)^2 \qquad (10.6)$$

10.4 RESULTS AND DISCUSSION

The proposed techniques were tested on CNTFET technology for the NAND gate. Simulations were conducted using a 32-nm technology file with a

Table 10.2 Comparative analysis of different technologies in term of performance parameters of existing circuits

Gates		MOS	FinFET	CNTFET
NOT Gates	Dynamic Power (μW)	0.176	0.131	0.009
	Delay (pS)	7.985	4.334	0.925
	PDP (aJ)	1.405	0.567	0.008
NOR Gates	Dynamic Power (μW)	0.203	0.118	0.003
	Delay (pS)	10.46	5.250	1.254
	PDP (aJ)	2.123	0.619	0.003
AND Gates	Dynamic Power (μW)	0.339	0.196	0.023
	Delay (pS)	15.40	4.857	2.234
	PDP (aJ)	5.220	0.951	0.051
NAND Gates	Dynamic Power (μW)	0.218	0.143	0.013
	Delay (pS)	10.32	5.713	2.134
	PDP (aJ)	2.249	0.816	0.027
EXOR Gates	Dynamic Power (μW)	0.226	0.136	0.043
	Delay (pS)	15.92	8.204	2.936
	PDP (aJ)	3.597	1.115	0.126
EXNOR Gates	Dynamic Power (μW)	0.204	0.142	0.013
	Delay (pS)	15.02	8.438	3.293
	PDP (aJ)	3.064	1.198	0.042

variation of supply voltage from 0.9V at 270°C temperature. The dynamic power dissipation (nW) was measured using transient analysis, which depends on the switching activity. For the measurements, a pulse width of 2 ns with a period of 4 ns, rise and fall time of 10 ps each, and a simulation time of 100 ns were used. The proposed techniques resulted in lower power dissipation due to reduced switching activity compared to other techniques. The simulation results from Tables 10.2 and 10.3 showed that the proposed techniques saved a maximum of 54.06% and 58.91% when compared to the conventional NAND gate. The proposed techniques exhibited the lowest power dissipation as compared to other existing techniques.

10.4.1 Benchmark C17 (ISCAS'85)

The C17 benchmark is part of the ISCAS'85 family of circuits, consisting of six NAND logic gates labeled G1 to G6, as shown in Figure 10.6. This circuit has five primary inputs and two primary outputs [14].

Table 10.4 displays the results of simulating the C17 circuit (ISCAS'85) at temperatures of 25°C and 110°C. The C17 circuit, along with B01 (ITC'99) and B02 (ITC'99), is a five-input NAND gate circuit that allows for a total of 32 input combinations. However, only the results for all 0's and all 1's are

Table 10.3 Comparative analysis of different technologies in term of performance parameters in proposed circuits

Gates		MOS	FinFET	CNTFET
NOT Gates	Dynamic Power (μW)	0.096	0.067	0.007
	Delay (pS)	5.584	2.434	0.725
	PDP (aJ)	0.536	0.163	0.005
NOR Gates	Dynamic Power (μW)	0.144	0.078	0.002
	Delay (pS)	8.64	3.562	1.152
	PDP (aJ)	1.244	0.277	0.002
AND Gates	Dynamic Power (μW)	0.231	0.107	0.019
	Delay (pS)	8.38	2.672	1.234
	PDP (aJ)	1.935	0.285	0.023
NAND Gates	Dynamic Power (μW)	0.134	0.083	0.010
	Delay (pS)	7.54	4.673	1.123
	PDP (aJ)	1.010	0.387	0.011
EXOR Gates	Dynamic Power (μW)	0.138	0.084	0.032
	Delay (pS)	12.05	8.204	1.736
	PDP (aJ)	1.662	0.689	0.055
EXNOR Gates	Dynamic Power (μW)	0.103	0.132	0.008
	Delay (pS)	11.04	6.182	2.162
	PDP (aJ)	1.137	0.816	0.017

shown. The low and high input states cover all worst-case scenarios for leakage power dissipation in the intermediate state of inputs, as the maximum number of OFF transistors occur when all the inputs are low and the minimum number of OFF transistors occur when all the inputs are high. Figure 10.7 shows the power-delay product (PDP) of the proposed design for NCFET and PCFET transistors.

Simulation results for carbon nanotube field-effect transistors (CNTFETs) with 1, 4, 8, and 16 tubes indicate that a 3% variation in the width of the channel feature (WFF) from its mean value leads to a significant deviation in IOFF current. Specifically, the deviation is nearly 110% for fin1 and 158% for tubes 4, 8, and 16. This represents the highest average standard deviation of I_{OFF}. Additionally, as the number of fins varies, along with other parameters, Figure 10.12 shows that a 10% deviation from the mean value of length results in a 42% deviation in width, 38% in height, 115% in WFF, 36% in Tfin, and 32% in N. These deviations from the mean also result in an average standard deviation in I_{OFF}, which is higher than the deviation observed for I_{ON}. The maximum I_{OFF} deviation is achieved for WFF variation. However, I_{OFF} deviations decrease as the number of fins increases.

Table 10.4 Simulation results of bench mark circuit at 32 nm using RLC interconnects

Bench mark circuits	Temp	Input vector	Leakage power dissipation (nW)							
			NAND gate		LECTOR NAND gate		Galeor NAND gate		Proposed NAND gate	
			MOS	CNTFET	MOS	CNTFET	MOS	CNTFET	MOS	CNTFET
C17 (ISCAS 85)	25°C	00000	64.634	49.42	28.124	8.214	4.2920	3.215	1.1457	1.023
		11111	51.639	41.92	46.318	32.184	27.700	24.863	1.4283	1.281
	110°C	00000	1724.1	1573.2	128.56	89.826	73.526	63.167	22.445	18.356
		11111	1213.1	1042.1	498.12	568.98	399.85	365.23	28.687	23.176
B01 (ITC'99)	25°C	00000	363.37	329.12	323.37	307.34	283.82	159.38	283.82	131.34
		11111	378.28	355.43	348.28	314.23	319.94	287.42	319.94	267.19
	110°C	00000	2718.3	2427.1	2569.1	2234.8	2318.3	2093.1	2318.3	2043.4
		11111	2531.3	2124.4	2246.4	1926.7	2002.4	1778.2	2002.4	1729.3
B02 (ITC'99)	25°C	00000	178.32	152.92	153.23	130.33	133.18	117.39	133.18	110.37
		11111	189.39	165.29	162.52	144.92	140.01	121.34	140.01	115.22
	110°C	00000	2238.4	1984.2	2017.3	1792.6	1826.2	1593.2	1826.2	1689.7
		11111	1935.8	1727.6	1736.1	1539.5	1539.9	1323.5	1539.9	1378.4

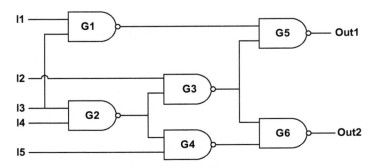

Figure 10.10 Benchmark C17 (ISCAS'85).

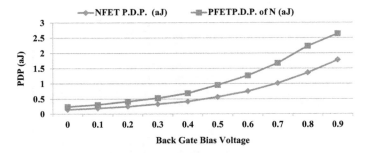

Figure 10.11 PDP of proposed NCFET and PCFET when reverse biased.

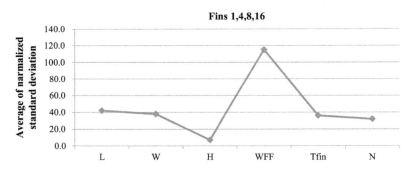

Figure 10.12 Multiple CNT tubes impact on I_{OFF}.

10.5 CONCLUSION

Several techniques have been proposed to reduce power consumption in CNTFET circuits. One of these techniques involves adding an extra transistor in the logic circuits as a drain gating or pass transistor logic. Another technique utilizes RLC interconnects. The proposed designs using these techniques were found to significantly reduce leakage power. For example, the proposed technique circuit showed a maximum saving of leakage power at 250°C of 98.36% for a 11 input vector and 99.83% for a 01 input vector. Similarly, the proposed technique circuit showed a maximum saving of leakage power at 250°C of 97.17% for a 01 input vector, 95.10% for a 10 input vector, and 94.57% for a 00 input vector. In SG mode, the proposed technique showed a maximum saving of leakage power at 250°C of 99.54% for a 00 input vector, and in LP mode, it showed a maximum saving of leakage power at 250°C of 99.90% for a 01 input vector. These results demonstrate that the proposed techniques can achieve significant savings in leakage power, with a maximum of 22.8% at 00, 49.45% at 01, 39.67% at 10, and 56.22% at 11 input vector, when compared with a basic NAND gate.

REFERENCES

[1] Y. Ye, S. Borkar, and V. De, A new technique for standby leakage reduction in high performance circuits. In: *IEEE Symposium on VLSI Circuits*, pp. 11–13, 1998.

[2] L. Wei, Z. Chen, M. Johnson, K. Roy, Y. Ye, and V. De, Design and optimization of dual threshold circuits for low voltage low power applications. *IEEE Transactions on Very Large Scale Integration*, vol. 7, no. 1, pp. 16–24, 1999.

[3] N. Sirisantana, L. Wei, and K. Roy, High-performance low-power CMOS circuits using multiple channel length and multiple oxide thickness. In: *Proceedings International Conference on Computer Design*, pp. 227–232, 2000.

[4] T. Karnik, et al. Total power optimization by simultaneous dual-Vt allocation and device sizing in high-performance microprocessors. In: *ACM/IEEE Design Automation Conference, DAC-02*, pp. 486–491, 2002.

[5] K. Roy, S. Mukhopadhyay, and H. Mahmoodi-Meimand, Leakage tolerant mechanisms and leakage reduction techniques in deep-submicron CMOS circuits, *Proceedings of the IEEE*, vol. 91, pp. 305–327, 2003.

[6] S. Mukhopadhyay, C. Neau, T. Cakici, A. Agarwal, C.H. Kim, and K. Roy, Gate leakage reduction for scaled devices using transistor stacking, *IEEE Transactions on Very Large Scale Integration (VLSI) Systems*, vol. 11, no. 4, pp. 716–730, 2003.

[7] J.C. Park, V.J. Mooney, and P. Pfeiffenberger, Sleepy stack reduction in leakage power. In: *Proceedings of the International Workshop on Power and Timing Modeling, Optimization and Simulation (PATMOS 04)*, pp. 148–158, 2004.

[8] N. Hanchate, and N. Ranganathan, LECTOR: A technique for leakage reduction in CMOS circuit. *IEEE Transactions on VLSI Systems*, vol. 12, no. 2, pp. 196–205, 2004.

[9] P. Verma, and R.A. Mishra, Leakage power and delay analysis of LECTOR based CMOS circuits, *International Conference on Computer & Communication Technology (ICCCT) IEEE*, pp. 260–264, 2011.

[10] S.A. Tawfika, and V. Kursun, FinFET domino logic with independent gate keepers, *Microelectronic Journal*, vol. 40, pp. 1531–1540, 2009.

[11] P. Mishra, A. Muttreja, and N.K. Jha, FinFET Circuit Design, SPRINGER, *Nanoelectronic Circuit Design*, pp. 23–53, 2011. doi:10.1007/978-1-4419-7609-3_2

[12] L. Nan, C. Xiao Xin, L. Kai, M.A. Kai Sheng, D. Wu, W. Wei, M.A. Rui, and Y.U. Dun Shan, Ultra-low power dissipation of improved complementary pass-transistor adiabatic logic circuits based on FinFETs, *Science China*, vol. 57, pp. 1–13, 2014.

[13] C. Meinhardt, A.L. Zimpeck, and R.A.L. Reis, Predictive evaluation of electrical characteristics of sub-22 nm FinFET technologies under device geometry variation, *Microelectronic Reliability*, vol. 54, pp. 2319–2324, 2014.

[14] L. Nan, C. Xiao Xin, L. Kai, M.A. Kai Sheng, D. Wu, W. Wei, M.A. Rui, and Y.U. Dun Shan, "Low power adiabatic logic based on FinFETs" *Science China*, vol. 57, pp. 1–13, 2014.

Index

For Product Safety Concerns and Information please contact our EU
representative GPSR@taylorandfrancis.com
Taylor & Francis Verlag GmbH, Kaufingerstraße 24, 80331 München, Germany

www.ingramcontent.com/pod-product-compliance
Ingram Content Group UK Ltd.
Pitfield, Milton Keynes, MK11 3LW, UK
UKHW021122180425
457613UK00005B/189